STARSAILING

THE WILEY SCIENCE EDITIONS

The Search for Extraterrestrial Intelligence, by Thomas R. McDonough

Seven Ideas That Shook the Universe, by Bryon D. Anderson and Nathan Spielberg

The Naturalist's Year, by Scott Camazine

The Urban Naturalist, by Steven D. Garber

Space: The Next Twenty-Five Years, by Thomas R. McDonough

The Body in Time, by Kenneth Jon Rose

Clouds in a Glass of Beer, by Craig Bohren

The Complete Book of Holograms, by Joseph Kasper and Steven Feller

The Scientific Companion, by Cesare Emiliani

STARSAILING

SOLAR
SAILS
AND
INTERSTELLAR
TRAVEL

Louis Friedman

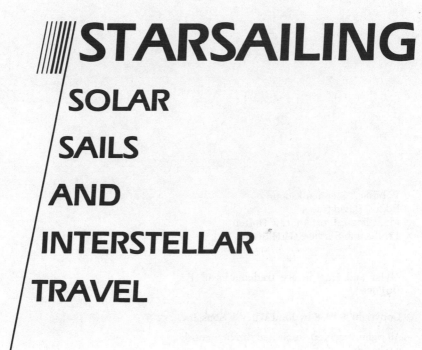

WILEY SCIENCE EDITIONS

John Wiley & Sons, Inc.
New York • Chichester • Brisbane • Toronto • Singapore

Publisher: Stephen Kippur
Editor: David Sobel
Managing Editor: Andrew Hoffer
Production Service: G&H SOHO, Ltd.

Mylar and Kapton are trademarks of E.I.
DuPont.

Library of Congress Cataloging-in-Publication Data

Friedman, Louis.
  Starsailing: solar sails and interstellar travel.

  (Wiley science editions)
  1. Solar sails.   2. Interplanetary voyages.
3. Astronautics in astronomy.   I. Title.   II. Series.
TL783.9.F75   1987   629.47'54   87-14229
ISBN 0-471-62593-0

Printed in the United States of America

88 89 10 9 8 7 6 5 4 3 2 1

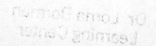

To Connie, my wife, companion and friend
through all of the excitement
at work and at home.

///// ACKNOWLEDGMENTS

I thank Carl Sagan and Bruce Murray for their encouragement, guidance, and support, and acknowledge their role in leading efforts for a positive future. Dr. Murray, as Director of JPL, led the effort to open minds and consider the solar sail for the Halley rendezvous. I wish to acknowledge the Solar Sail team at JPL, especially William Layman, Jerome Wright, and the late William Carroll for much of the technical work described herein. Dr. Ken Atkins, who helped in getting the sail project started at JPL and led the Ion Drive effort as its competitor, deserves acknowledgment as one of our innovators. I wish also to thank Jim Gilbert for suggesting the book and David Sobel, editor at John Wiley and Sons, for his support and patience, especially in dealing with an author who always seemed to have something else on his mind. Charlene Anderson, editor of *The Planetary Report*, and her assistant Donna Stevens deserve special recognition for their hard work in helping make this a whole book instead of a sum of parts. Thank you. I also thank Lorna Griffith and Donna Lathrop, for the typing of the manuscript.

A Mars base. The great advantage of a solar sail spacecraft is that it can carry enormous payloads to *and from* the planets without consuming any fuel. It is ideal for regular trips back and forth between Earth and its inner-planet neighbors. Mars is the next major exploratory step for the human species. Astronauts and cosmonauts will begin human exploration in the twenty-first century. A logical way to support humans on Mars will be to set up logistical support via solar sail missions. The figure shows a solar sail spacecraft in orbit over Mars, having carried and delivered surface vehicles for exploring, roving, and building on Mars. The ladder on the main vehicle reminds us this will be a human-occupied base. (Ken Hodges, NASA illustration)

||||| CONTENTS

STARSAILING

///// INTRODUCTION

Sailing is a romantic idea. The word itself conjures up images of children sitting on the shore, staring out to sea at ships passing over the horizon to far-off adventures in barely imaginable places. Sailing is people in harmony with nature, feeling the wind and the sea in their faces as they struggle to harness the forces of nature for great adventure and discovery. It is a peaceful notion of riding along free of any mechanical help.

Solar sailing provides similar images: Peter Pan catching a moonbeam to sail upon, Icarus flying to the sun with outspread wings. It will allow us to fly in space—free of propellant and of mechanical assistance.

We have only begun to travel into the solar system. Our journeys are limited by the great length of time it takes to travel between the planets and by the small size of the payloads that conventional rockets can carry. The U.S. shuttle and the Soviet Proton, for example, can deliver only 5 tons to Mars or 2 tons to Jupiter, and that simply isn't enough. For advanced robotic exploration of those planets, we must send heavy roving vehicles to land and explore the surfaces and special equipment designed to collect samples and bring them back to Earth. Conventional ballistic rockets are also woefully inadequate for transporting humans, since manned missions would mean having to carry enough fuel for the return journey. To solve these problems we need to find ways to travel in space with less fuel. One possible solution is to develop the solar sail, which uses no fuel at all.

How realistic is solar sailing? Three times NASA has rejected it as a means of propulsion for spacecraft. Does that mean it is an idea whose time has not yet come? Or has it come and passed? There are a number of practical difficulties with solar sailing, and it may well be that the idea of using light sails for free transportation around the solar system may never come to pass. But even if the solar sail remains only a dream, it will force us to consider new ways to explore our vast solar system, and if it does that, it will have been enormously important.

However, some of us believe that solar sailing is a practical idea. We believe that a solar sailing vehicle could travel about the solar system, moving humans and machines, instruments and supplies for scientific, technological, and cultural exploits in space. We understand the engineering problems, and we have come up with workable, affordable solutions.

Solar sailing is not the only idea for advanced propulsion, of course. Among the others are bigger chemical rockets, lighter or faster-burning fuel, electric propulsion that uses solar- or nuclear-power sources, assembly of rockets in space for multiple stages, and in the more distant future, fusion power. Laser propulsion is another possibility: It is an advanced variation of solar sailing in which light from powerful lasers will replace the energy provided by sunlight. We will discuss all of these types of propulsion later.

This book, then, is an examination of solar sailing—what it is, how it's done, and what it's good for. In it, I outline the history of solar sailing, and explain what solar sails are, how they work, and why they offer such promise for the future of travel throughout the solar system. I also discuss the physics of solar sailing, the design and construction of solar sail spacecraft, and the problems of solar navigation. I've taken care to present this information in nontechnical language, with no intent to provide a detailed engineering analysis for solar sailing. In the matter of units—meters, feet, kilograms, and so on—I do not attempt to be consistent. My goal is to make information about the solar sail available to many.

I often compare sailing in deep space to sailing on the high seas, not because the physical principles are the same—they are not—but because the comparisons can help us understand the techniques of solar sailing and the role it could play in space exploration. Like terrestrial explorers, space explorers are driven by romantic ideas, curiosity, yearning, and the desire to gain scientific and utilitarian advantages. In other words, we want to explore the solar system for many of the same reasons our predecessors chose to explore this planet.

The aim of many popular books on terrestrial sailing is to motivate people to go out and sail or to begin building their own sailboats. Naturally, I cannot expect my readers to take up solar sailing physically, but I do hope some of you will take up the sport conceptually, that you will in some way be a part of the exploration of space via the solar sail.

The basis for this book was a $4-million solar sail team effort at the Jet Propulsion Laboratory (JPL) in California in 1977 and 1978. The purpose of that effort was to design a vehicle that would permit an in-depth, up-close examination of Halley's Comet.

I am indebted to many people who worked with me on that team effort, some of whom are mentioned in the text. I especially want to acknowledge the help of Jerome Wright, the developer of the solar sail technique and the discoverer of the Halley's Comet opportunity; William Layman, the chief engineer of our team; and William Carroll, the materials-team leader. Layman once asked, "Isn't it strange that everywhere we go—at NASA, at universities, in industry—solar sailing seems to attract the best people?"

It is strange indeed, but it is also true. I have been privileged to work with some of those people. On the Halley's Comet project, those "best people" were driven by their understanding of what the solar sail could do. Resisted by the bureaucracy, they raced against the unrealistic deadline and learned to make rapid, high-risk decisions. Success remained out of reach for the sail team, because the obstacles were too great; but none of us regretted the effort—and management never quite forgave us for trying.

The exploration of the solar system seems destined to be one of the most valuable legacies of our generation. Few, if any, other enterprises of humankind are so certainly lasting, benign, and creative. Few express as much optimism. This is one of the reasons that I joined with Carl Sagan and Bruce Murray to form The Planetary Society in 1979. No other two people on the planet had a clearer sense of the desire of people all over the world to take part in space exploration, or a more profound sense of loss at seeing the U.S. program withdraw in a misguided redirection of priorities and fear of the future. Solar sailing, Halley's Comet missions, the search for extraterrestrial intelligence, the continued exploration of Mars, and missions to the edge of the solar system were all dropped from U.S. plans within a three-year period. Reversing this trend is the aim of The Planetary Society—now the largest space interest group in the world with more than 100,000 members.

1

THE HISTORY OF SOLAR SAILING

Sketch by Orban from an early publication on solar sailing, "Clipper Ships of Space,"
by Carl Wiley, 1951. (Reprinted from *Astounding Science Fiction*)

At a meeting of the Advanced Projects Group at the Jet Propulsion Laboratory, Chauncey Uphoff and Phillip Roberts told me of some work on solar sailing being done by Jerome Wright, an engineer at the Battelle Memorial Institute. The actual concept of solar sailing was pretty new to me at that time, and I didn't know quite what they were talking about—"propelling" a spacecraft by means of the force from sunlight pressure. Of course, I knew about sunlight pressure and, in fact, had done some work analyzing it in my thesis years earlier. Sunlight pressure is one of the important perturbation forces that affect spacecraft motion. Very early in the space age, the National Aeronautics and Space Administration (NASA) had launched small metal needles in the ionosphere to study communication and the way signals behave in the ionosphere. A major controversy had greeted the prediction that the sunlight pressure would push these needles in such a way that their orbit would be lowered and they would harmlessly burn up in Earth's atmosphere. When they did exactly that, fulfilling the prediction, the force of solar pressure was confirmed as was the magnitude of the solar constant.

I also knew that sunlight pressure had been a significant force in controlling the attitude, or orientation, of the *Mariner 10* spacecraft that went to Venus and Mercury. Aboard that spacecraft were some small panels that acted as vanes when they were tilted at different angles to the sun. As a result of the sunlight pressure hitting those vanes, the spacecraft could be made to turn on its axis. This is called an *attitude force,* and it causes a rotational motion—in other words, it turns the spacecraft but does not put it on a different trajectory. A *translational force* changes the trajectory, and trajectory calculations for interplanetary spacecraft must take into account the perturbation from sunlight pressure on the motion of the spacecraft.

The actual use of sunlight pressure as a means of propulsion was a novel idea, although the theory that it *could* be so used had been around for many years, having been suggested back in the 1920s. Still, it didn't sound very practical. We already had a means of low-thrust propulsion—solar-electric propulsion—on the drawing boards, and there were many proponents of an advanced low-thrust system using nuclear-electric propulsion. These two methods appeared to be adequate to all the propulsion requirements we could think of for interplanetary travel. Why get interested in another academic technique?

Then, at the group meeting, Chauncey Uphoff and Phillip Roberts dropped the bombshell. "Jerry Wright has found a possible way to rendezvous with Halley's Comet," Chauncey announced.

"You mean fly-by, not rendezvous," I said.

"No, I mean rendezvous."

"With a trip time of ten years?"

"Would you believe four years?"

Prior to this discussion, as far as we knew, the only known way to rendezvous with Halley's Comet was a rather theoretical design. This design called for a very advanced solar- or nuclear-electric propulsion system that could find trajectories that could result in a rendezvous with the comet about ten years after launch. *Rendezvous,* by the way, means not just passing by the target but matching its speed and actually spending some time traveling with it. In other words, to rendezvous with another celestial body, the spacecraft must first get to the target and then match its orbit perfectly and fly in formation with it.

• Any rendezvous in space is tricky, but a meeting with Halley's Comet is especially difficult, because it requires us to "stop the world and get off." Here's why: Earth goes around the sun at a speed of approximately 30 kilometers per second. Halley's speed around the sun is, on the average, slower, since it is on a much larger ellipse. But when the comet gets near the inner solar system it is on the fast part of the ellipse. Kepler's laws of elliptical motion state that equal areas are swept out in equal times. For an elongated ellipse this requires the speed near the sun to be much faster than those in the outer solar system. As a result, the speed of the comet as it nears Earth's orbit is more like 40 kilometers per second. But the comet goes backward through the solar system—that is, it moves clockwise around the sun while Earth and the other planets move counterclockwise. And because the comet is going the other way, its relative speed is about 70 kilometers per second. This means that a spacecraft that wants to rendezvous with the comet has to lose all of its motion around the sun and then begin going in the other direction. Such a rendezvous had seemed too challenging—until Chauncey Uphoff spoke up at the meeting, and suddenly it was possible. We wanted to hear more.

Chauncey Uphoff's first step was to write a memo in which he explained the principles of solar sailing. The mission designers at the Jet Propulsion Laboratory (JPL) were so interested in the subject that in the spring of 1975 we invited Jerome Wright to come out from Battelle and conduct a seminar on the concept. He gave the seminar in May. This proved to be such a success that we persuaded Wright to come to JPL to work on the solar sailing idea and study the practical possibilities for a rendezvous with Halley's Comet. Wright arrived in December 1975; NASA approved a small study at JPL; and a team began to study the design of a solar sailing vehicle and the possibilities of a sail mission.

Although the idea of the solar sail had come as something of a surprise to us, the concept has a rich technical history dating back to the 1920s. The JPL study team, especially Uphoff and Wright, ferreted out most of the early literature on solar sailing.

The first scientists to mention the use of solar pressure as a propulsive force were Russians. One was the great space pioneer Konstantin Tsiolkovsky, and the other was an engineer, Fridrickh Arturovich Tsander, who in 1924 wrote, "For flight in interplanetary space I am working on the idea of flying, using tremendous mirrors of very thin sheets, capable of achieving favorable results." It is noteworthy that in the same article Tsander proposed Earth-orbiting space stations. Scientists at that time knew that light exerted pressure because the theory of electromagnetism, developed in the 1860s by James Clerk Maxwell, had been proved in various physical experiments in the nineteenth century. Maxwell himself provided the basic description of light as a "packet" of energy acting as if it were made up of tiny atomlike particles. These particles are called photons, and like other particles, they obey physical laws of motion. Photons have energy and momentum when they move, and the "sailing" is made possible by the transfer of momentum when a photon bounces off the reflective sail.

The first serious technical paper about solar sailing for spacecraft propulsion was Carl Wiley's "Clipper Ships of Space," a nonfiction article published in the May 1951 issue of *Astounding Science Fiction*. It is not surprising that solar sailing has often been mentioned in science fiction. Various twentieth-century stories include suggestions for interplanetary and interstellar travel on solar sails. Wiley, an aeronautical engineer, published the article under the pseudonym "Russell Sanders" because he didn't want to lose scientific credibility by suggesting "way-out" ideas in a science-fiction magazine. Many years later, however, during the JPL study—when the subject was no longer considered "way out"—Wiley, then an engineer at Rockwell, came to some of the technical presentations. Even in 1951, in fact, the topic was taken seriously enough that, in the same issue of the magazine, the space writer Willy Ley commented on Wiley's article. Ley liked the idea of solar sailing but didn't think it practical at that time. He thought it couldn't happen until "after rockets opened up space and enabled us to build artificial satellites" because too much time would be spent on slow trajectories spiraling away from Earth and to another planet.

The first article in a professional publication was written in 1958 by Richard Garwin, a Defense Department consultant with IBM. Dr. Garwin's paper was published in *Jet Propulsion* and included preliminary calculations of sail-vehicle performance. Following that, a number of technical papers were presented in the engineering literature and in NASA and university publications. Later, in the mid-1970s, Garwin was partly responsible for NASA's renewed interest in solar sailing. NASA administrator James Fletcher commissioned new studies after communicating with Garwin about the concept. Those studies were then assigned to Jerome Wright at Battelle Memorial Institute in Ohio.

Also in 1958, Ted Cotter, at the Los Alamos Scientific Laboratory, came up with the notion of spinning the sail. Cotter also wrote a short article describing the technical aspects of sailing in the early 1960s. *Time* magazine picked up on Cotter's work in a 1958 editorial, "Trade Winds in Space."

In 1960, Philippe Villers wrote his master's thesis on the subject at Massachusetts Institute of Technology (MIT). A meeting on solar sail design was held that same year at NASA's Langley Research Center, and a short course on solar sailing was offered at the University of California at Los Angeles (UCLA) the following year. At first, scientists concentrated on spinning sails (an example is shown in Figure 1.1), although they also considered rigid sails, especially for spacecraft stabilization. Between 1965 and 1967, Richard MacNeal and John Hedgepath invented the heliogyro, a spinning helicopterlike vehicle.

NASA began technology studies in the mid-1960s. These studies examined various designs and technology requirements for solar sailing vehicles without reference to specific missions. As the space program began to shrink after the *Apollo* missions, however, NASA dropped this work. By the mid-1970s, no research at all was going on in solar sailing except for the small study by Jerome Wright at Battelle.

Wright did his work under a contract to NASA. This contract had as its primary purpose the calculation of a launch vehicle and the propulsion requirements for various missions that were then under consideration. Because he wanted to be thorough, he also did a cursory analysis of the possibility of using solar sail capabilities for traveling to other planets in the solar system.

Then Wright found the Halley rendezvous opportunity. By the time the JPL study team began its task, two major developments had occurred. First, NASA was developing the space

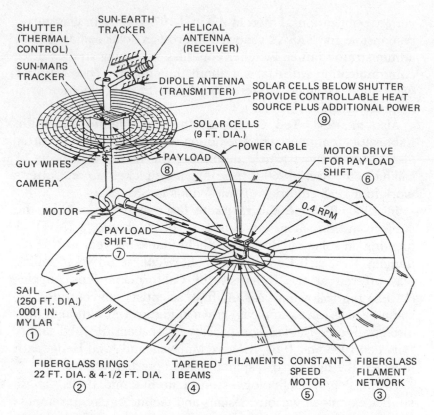

SHUTTER (THERMAL CONTROL)
SUN-EARTH TRACKER
HELICAL ANTENNA (RECEIVER)
SUN-MARS TRACKER
DIPOLE ANTENNA (TRANSMITTER)
SOLAR CELLS BELOW SHUTTER PROVIDE CONTROLLABLE HEAT SOURCE PLUS ADDITIONAL POWER ⑨
SOLAR CELLS (9 FT. DIA.)
POWER CABLE
MOTOR DRIVE FOR PAYLOAD SHIFT ⑥
GUY WIRES
PAYLOAD ⑧
CAMERA
MOTOR
0.4 RPM
PAYLOAD SHIFT ⑦
SAIL (250 FT. DIA.) .0001 IN. MYLAR ①
FIBERGLASS RINGS 22 FT. DIA. & 4-1/2 FT. DIA. ②
TAPERED I BEAMS ④
FILAMENTS
CONSTANT SPEED MOTOR ⑤
FIBERGLASS FILAMENT NETWORK ③

FIGURE 1.1 Spinning sails were considered first because the spin provides a means of stabilizing the sail without a structure. The mechanisms for controlling the vehicle are, however, rather complicated. This is a concept for a small Mars vehicle. (Philippe Villers, MIT, American Rocket Society preprint, December, 1960)

shuttle, which promised to carry large-volume payloads into orbit. Second, there had been great advancements in the technology of deploying huge structures in space. The shuttle also made it possible for scientists to test space concepts, and the JPL study team hoped to test the solar sail from a shuttle in orbit.

The positive results of the 1976 and early 1977 JPL studies captured the imagination of the new director of the Jet Propulsion Laboratory, Dr. Bruce Murray. With his approval, the study team made a major effort to put together a project plan for a rendezvous with the comet. This work, however, had to be done rapidly. In order to launch in late 1981, the project would have had to start moving by the end of 1978. I was put in charge of the study and we quickly wrote a proposal for a one-year study and gave it to NASA.

NASA accepted the proposal, and in 1977–78, the JPL team conducted a solar sail design study for the mission, with the help of a half-dozen industrial contractors and support from NASA's Ames and Langley research centers. The year-long work on preliminary design demonstrated that, indeed, solar sailing was a feasible spacecraft-propulsion technique.

Despite the confidence of the technical team and the completion of a valid preliminary design, however, the NASA management was conservative. They felt the design and implementation could not be accomplished in time for a 1981 launch to Halley's Comet. NASA also thought that the technology for solar sailing was not sufficiently "mature" to be implemented on a near-term space project. Indeed, the Halley mission requirements were severe—and even our willingness to incur great risk for great gain was insufficient to overcome management's skepticism. And as it turned out, the conservatives were right, we could not have done it. It was a self-fulfilling prophesy. In general, our space abilities were slowing down. NASA assumed that the space shuttle would be operating and ready for interplanetary flight by 1981. This assumption was even more optimistic than the ones we had made about solar sailing.

Also, the sail proposal had become caught up in a competition with a proposal by solar-electric propulsion advocates for a complex advanced low-thrust ion-drive system. (Ion drive is a propellant-driven system that provides rocket propulsion by expelling ions continuously from on-board thrusters. The ions are made by converting solar- or nuclear-electrical power, in high-energy vacuum chambers, a process that strips away electrons from a gas such as mercury or argon.) Although solar-electric propulsion won the competition, it lost the war: It was quickly rejected for the Halley mission, and the United States ended up with no comet mission at all. At this point, NASA withdrew its support for work on solar sailing, although some smaller research programs continue.

The solar sailing idea, however, had captured the imagination of workers involved in space mission design and planning around the world. A group in France had begun work on a design for *la voile solaire*. Another private group in France now carries on some work in solar sailing. After the NASA program was terminated, a group of engineers, principally from JPL, set up a private organization partially to seek public funding in order to carry on solar sail development. This group, the World Space Foundation, has actually fabricated a prototype sail and conducted a test deployment demonstrating how it would work. The private efforts to develop solar sailing are discussed in Chapter 10.

//////2
PRINCIPLES
OF
SAILING

The solar sail works like a mirror. The reflected sunlight provides a force to push the sail. The bigger the sail, the greater the force—a half-mile square sail like that designed for the Halley Comet mission in low Earth orbit would be visible from Earth—even in daylight. This solar sail designed by the World Space Foundation shows the Earth reflected on the aluminized surface. (Painting by Tom Hames/World Space Foundation)

The solar wind plays no part in solar sailing; it's important to understand this from the start. The solar sail operates on sunlight pressure—the pressure produced by light when it "bounces off" a mirror. This force is 1,000 to 10,000 times greater than that of the solar wind, and while there are many analogs between terrestrial sailing and solar sailing, we must disabuse ourselves of the idea that the solar wind can propel us through space.

Let's clear up any confusion between these two phenomena. What, first of all, is solar wind? Solar wind is made up of high-velocity electrons and protons emitted by the sun as it burns up hydrogen. These particles stream through the solar system, away from the sun, and produce various effects when they hit spacecraft or planets. One of the most notable effects occurs when solar-wind particles hit Earth's ionosphere and cause magnetic storms, which interfere with our communications. However, the energy from these particles is much smaller than the energy in the light particles that constitute solar pressure. These light particles, or photons, are part of the sun's electromagnetic radiation output. They are bundles or packets of energy. In reality, photons have no mass, but because they act like particles it is useful to think of them as such. When photons hit something, they carry momentum, and they bounce off according to the laws of optics. (Strictly speaking, photons have no "rest" mass, i.e., mass when there is no motion; but they do have mass when moving. The mass is significant at relativistic speeds—speeds near the speed of light.)

Light is part of what we call the electromagnetic spectrum—*electromagnetic* because it is created by the interaction of electric and magnetic fields and particles, and *spectrum* because each interaction happens at a particular frequency. Some interactions are at very low frequencies, like the one that creates the electricity in your home (60 cycles per second); others are at very high frequencies, like those of radio and television broadcasts (for example, 100 megahertz—100 million cycles per second). Frequency is measured in cycles per second, called *hertz*. The interaction of electrons, protons, and other parts of the atom gives the atom energy. At each energy level the protons emit radiation at a particular frequency. Over all energy levels existing in nature there is a continuous range of emitted frequencies. This is the electromagnetic spectrum. Visible light has frequencies in the range of 700 to 1,000 trillion hertz. This may be considered the frequency of photon emission, photons being the "particles" of light.

The solar sail is a highly reflective "mirror." When photons hit this mirror, they impart a force to it and, hence, to the sail vehicle. If the sail is small, it gets only a small amount of force. But lots more photons hit large sails and give them a greater force. By tilting the mirror, or sail, in different directions, we can direct the force wherever we choose. In this way we can steer the vehicle through the solar system.

A terrestrial sailboat operates not just because of the wind hitting the sail but also because of the combination of the water and the wind. In other words, the combination of the two media—wind and water—causes the boat to move in a certain direction and at a certain speed. If it weren't for the rudder in the water, the wind would merely push the sail in a straight line, and the boat would be uncontrollable.

Similarly, a combination of two media causes the solar sail vehicle to move. In solar sailing, however, the two media are *sunlight photons* and *orbital velocity*. The sunlight photons are the space equivalent of wind in that they propel the vehicle through space. Orbital velocity—the speed at which the vehicle moves through its orbit—is the space equivalent of water, since it is the medium on which the solar sail vehicle travels.

All objects in the solar system move about the sun in elliptical orbits. This was first discovered by Johannes Kepler in 1609 and then explained by Isaac Newton in his law of gravitation in 1687. Kepler's laws of elliptical motion relate the speed of an object to its distance from the sun. Each orbit has a particular average speed. If the orbit is circular, the speed is constant. Most orbits are not circular, though, and so the speed varies in different segments of the orbit. Mercury, the planet closest to the sun, moves through its orbit faster than any other object in the solar system (88 days for one revolution around the sun); Venus is slower (225 days); and Earth is still slower (365¼ days). When we get all the way out to Pluto, the distant planet, the movement is very slow indeed (248 years). Earth travels approximately 940 million kilometers (the circumference of our ellipse) around the sun in one year at an average speed of approximately 30 kilometers per second. The average speed of Mars is slower, since it's farther out—about 24 kilometers per second. To move anything from the orbit of Earth to that of Mars, then, requires a change from the speed of Earth (30 kilometers per second), to that of Mars (24 kilometers per second), a change of about 6 kilometers per second (13,000 miles per hour, neglecting effects of inclinations and eccentricities of the orbit). To move about in the solar system, then, a vehicle has to be able to change its velocity and change the direction of its orbit relative to the sun.

In ordinary rocket propulsion we do this by firing an engine to produce a rocket thrust that changes our velocity almost instantly. The thrust also moves us outward from the sun, or inward toward it—outward if we direct that thrust behind us on the orbit to speed us up, inward if we direct the thrust ahead of us on the orbit to slow us down (a retro-maneuver). The velocity change to get us into Mars's orbit can be done in two burns: one as we leave Earth and another as we approach Mars. With a solar sail the procedure is different, but the principle is the same. We direct the sail so that a net force is produced from the photons bouncing off the sail—along the orbit to speed us up, or against the orbit to slow us down. In this way we change the orbit's velocity continuously without using any form of propellant. This is the inherent beauty and power of sailing—we don't need fuel to propel us; we simply use nature to propel us along. This also answers the question most often asked of me—in both technical and public lectures—about solar sailing: How can we tack (change direction and, in particular, go inward toward the sun)? The answer to that question iş: by changing the angle of the solar sail with respect to the sun's direction so that the sail catches the particles of light on a slant. This adjustment produces a force on the vehicle perpendicular to the sail. The sunlight pressure bounces off a solar sail at an angle, just as the wind bounces off a terrestrial sail, modifying the vehicle's velocity so as to produce changes in the orbital distance and direction that the spacecraft travels. If the velocity is increased we go outward. If it is decreased—inward. Just as we could not tack on Earth if it weren't for the water in which the sailboat moves, neither could we tack in space if it weren't for the orbital velocity we had and needed to modify in order to change direction. Change of direction comes about by adding to, or subtracting from, the orbital velocity.

The basic geometry of solar sailing, in which photons from the sun hit the mirror in one of two ways and cause a force that either increases or decreases the speed of the vehicle, is shown in Figure 2.1. With the sail, the change in velocity is applied slowly but continuously, unlike the sudden thrust of a rocket. Therefore, the sail vehicle builds up speed gradually rather than in instantaneous bursts. Because the change in orbit is also continuous, the path that a solar sail follows looks more like a spiral than an ellipse. It is really a series of continuously changing ellipses.

At what speeds can the sail vehicle move through the solar system? To find this out we must first understand the magnitude of the propulsive force. The sun, of course, emits enormous amounts of energy, and the amount of energy that propels the solar sail is related to the power output of the sun. Most of the radiation of the sun comes from the visible light emitted from the 6,000°F solar surface. The energy of the sun is spread out. As a result of this spreading, as one goes farther from the sun the amount of energy per unit area decreases in proportion to the square of the distance from the sun.

The amount of energy per unit time, per unit area, that reaches Earth is called the *solar constant*. Energy per unit time is what we usually call power. In common units of power the

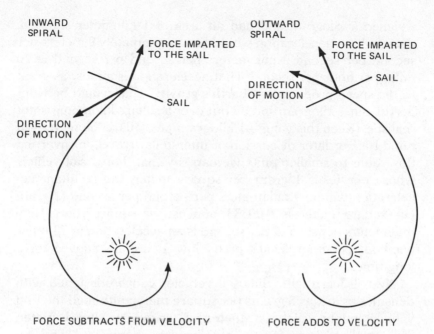

FIGURE 2.1 The principle of solar sailing: Sunlight bounces off the mirrorlike sail, imparting momentum to the vehicle. The force imparted is in a direction controlled by the angle of the sail with respect to the sunlight. It can either increase or decrease the velocity of the vehicle depending on the direction of the force.

solar constant is 1.4 kilowatts per square meter—that is, an area approximately 3 feet by 3 feet intercepts enough sunlight to keep a hand-held hairdryer running continuously. To find out how much force this produces on the sail and, hence, how much acceleration it can provide to the vehicle, you divide the power by the speed of the photons (the speed of light equals 300,000,000 meters per second) and multiply by 2 (which comes from the photons bouncing off the mirror).* If the sail vehicle

*Equations are mathematical sentences. This sentence is written $F = 2P/c$ where F is the force, P is power, and c is the speed of light. Since $F = ma$ (mass times acceleration), the acceleration $a = 2P/mc$. In the next sentence we substitute values in the equation: $m = 1$ kilogram, $P = 1{,}400$ watts times 1,000 by 1,000 square meters (the sail size), $c = 300{,}000{,}000$ meters per second. The answer is $a = 9.33$.

weighed a kilogram and had an area of 1 kilometer squared, the resultant acceleration would be approximately 9 meters per second per second. Nine meters per second per second is 20 miles per hour per second. That acceleration is almost as great as the speed produced by Earth's gravity. That would be wonderful, but 1 kilogram spread out over a square kilometer is not realistic. (Keep this value—1 kilogram per square kilometer—in mind for our later discussion of interstellar travel.) Converting this value to smaller units we can say that: For a sail vehicle whose density is 1 gram per square meter, the resultant acceleration will be 9 millimeters per second per second (1 gram per square meter is 0.00333 ounces per square foot). Nine millimeters per second per second is an acceleration of 72 miles per hour per hour. That's pretty low, unless it's applied for a long time.

We will learn later that sail vehicles can be designed with densities as low as 5 grams per square meter, although the first vehicles might weigh as much as 8 grams per square meter. This implies characteristic accelerations between 1 and 1.6 millimeters per second per second—that is, a change in speed from 0 to 8 or 13 miles per hour, in one hour. Not very much! But applied every second of every minute of every day over a space flight it can build up—a lot.

A characteristic acceleration of only 1 millimeter per second per second in one day builds up to a velocity change of nearly 100 meters per second (225 miles per hour) and a position change of 7,500,000 meters! In about 12 days we can have a velocity change of 1 kilometer per second and a position change of 90,000,000 meters. Thus, we see it is possible to build up the velocity change of 8 kilometers per second that is needed to go from Earth to Mars in about 100 days. That is a long time, however. In 100 days the Earth moves one-third of the way around its orbit, and Mars moves through one-sixth of its revolution. To shape the orbit properly for matching speeds with Mars, a velocity change must be made to take into account both speed difference and geometry. There is a constant need for changing the shape of the orbit so that the rendezvous with Mars is perfect. This means that the trip will take about

400 days. To get an idea of why this is so, imagine a group of terrestrial sailors who are five minutes away from the dock, as the crow flies, but who can't sail straight toward the dock without crashing into it, because they are sailing before a strong wind. To avoid a collision, they will take a roundabout path to the dock, tacking (or jibing) back and forth in order to slow themselves down. This will get them there safely but it will take a longer time. In solar sailing, the effect is more pronounced because the dock (Mars) is moving and our accelerations are very small.

Four hundred days is slower than most rocket-propelled flights to Mars. There is no performance advantage of low thrust (either sail or electric propulsion) for simple one-way trajectories to nearby planets. But low thrust, in particular the sail, has an advantage in more difficult missions, such as a rendezvous with a comet or a round-trip in which one-way speed is less important. The sail's advantage is that it allows the vehicle to carry large payloads. Rockets can get a spacecraft to Mars faster than a sail can, but the rockets have to use a lot of their energy in expelling fuel. The sail uses no fuel, and this—together with continuous thrust—enables it to carry more mass. We'll look at this more closely later on.

To sum up: Although the solar sail is a low-thrust or low-acceleration vehicle, the continuous energy from the sun allows it to move about the solar system with enormous velocity changes building up in a relatively short period of time, which gives the sail vehicle the ability to travel over great distances. The sail does not allow us to change direction quickly, nor does it provide huge bursts of acceleration. It is, however, an ideal way to move large masses over great distances in an efficient fashion.

One might think that it wouldn't make too much difference to the performance of the sail vehicle if the characteristic acceleration were 1 millimeter per second per second or 1.5 millimeters per second per second. However, because the acceleration directly translates into the mass-carrying capabilities of the payload, and especially the mass ratio of the payload compared to the mass of the entire sail vehicle, it turns out that that 50 percent makes an enormous difference in the size of the payload that the sail vehicle can carry. These values of acceleration are directly tied to the design of the sail vehicle (expressed by its density in grams per square meter). Thus, it is the difference in the design that causes the performance of the sail to be either ho-hum or fantastic.

Terrestrial sailors will find it easy to understand this situation: Small differences in the design of a terrestrial sail and in the weight of the vessel make an enormous difference in the speed and capabilities of a sailboat and in its performance.

Now that we have an understanding of the basic physics of a solar sail and an idea of how it works, we are ready to consider the designs of solar sail vehicles and potential uses for the sail.

3

THE DESIGN
AND
CONSTRUCTION
OF
SOLAR SAILS

The heliogyro solar sail presents the large reflected area to the sun in an array of very long "blades," like helicopter blades. The entire vehicle spins, providing stability. The blades can be pitched to provide pointing control. (NASA illustration)

To understand the basic components of a solar sailing ship, you need only two materials: aluminum foil and plastic wrap of the sort you use in the kitchen. The foil will be the basic constituent of the sail—a large, smooth, highly reflective sheet of thin, lightweight material. The plastic will serve as the strong, thin lining on which the reflecting material will be mounted.,

You need a very shiny mirror, or reflector, so that a great many photons will bounce off it, giving momentum to the sail vehicle. A perfect reflector is not possible; some photons will always be absorbed by the reflecting material. Silver would be an ideal reflector, except that it is very expensive and it oxidizes easily, and so most sails will be made of the next best material: aluminum, which is inexpensive and does not oxidize rapidly. "Reflectances" of 85 to 88 percent are possible with aluminum.

The sail must also be smooth, because in wrinkled areas the energy of reflected sunlight is wasted, and also because wrinkles can make the sunlight reflect back on the sail material, which could produce hot spots. Remember that light has energy, and if it is intense enough it can cause heating when it hits an object. For example, if we concentrate sunlight through a magnifying glass, we can burn a hole in paper or start a fire. The same thing happens when we focus light narrowly with a mirror. The sail is a mirror, and the wrinkles can focus the light. Wrinkles could even cause parts of the sail to burn. Thus, we need to hang the sail as smoothly as possible.

This will introduce some important design considerations for the wires and edge members that hold the sail in its proper position. Unroll some aluminum foil and see how difficult it is to keep it smooth. This is one of the reasons the sail material won't be pure aluminum. We need to underlay the aluminum with plastic so that it will be tough enough to survive folding and packing for launch and deployment in space. Another reason not to use pure aluminum is that the sail would be too heavy. We can make a lighter, smoother sail by spraying a very thin layer of aluminum on a plastic substrate. Someday, when manufacturing in space is possible, we may be able to build a thin sail of pure metal with no plastic substrate at all. But this is not possible as long as we have to package the sail, fold it, and then deploy it in space.

While Eric Drexler was a student at Massachusetts Institute of Technology, he developed a concept for space manufacturing of very thin solar sails. You might, for example, construct the sail on fiber strings like those in a spider web. The very thin aluminum could be evaporated in the vacuum of space so that it would form a film between the fiber strings on a soaplike bubble that would "disappear," leaving only the thin aluminum reflective coating. Drexler has shown that it would be possible to use this method to produce a sail about 0.1 micron (4-millionths of an inch) thick (the thickness of only a few hundred atoms). With such a sail, the acceleration would be 8 millimeters per second per second! This sail would be an extraordinary machine for traveling between the planets.

Now, if you read the label on the package of plastic wrap, you will note that this plastic is about 0.5 mil thick. One mil is 1 one-thousandth of an inch (0.001 inch). In metric units, that is 25.4 micrometers.* The usual word for micrometers is microns. Thus, ordinary plastic wrap is approximately 13 microns thick. A sheet of plastic wrap weighs approximately 16 grams per square meter. If you were to make a solar sail out of that material, your best acceleration would be a little more than half a millimeter per second per second. This is rather sluggish for traveling around the solar system, so we need to use something a little thinner than ordinary household plastic as the base on which we will spray the aluminum. The thinnest

---

*Physicists use many different units to measure atomic-size distances. One micron (micrometers) is one-millionth of a meter, or one-thousandth of a millimeter. It is 0.04 mil—that is, four-millionths of one inch. It is 10,000 Angstroms. An angstrom is approximately the radius of an atom.

usable metal is about 0.02 micron. Below that, photons would go through the metal instead of bouncing off. In our first sail, made on Earth and packaged for flight, we will use about five times the minimum—approximately 0.1 micron of vacuum-deposited aluminum—to achieve the necessary reflectance. This very thin layer of metal will add little weight to the sail.

Plastic film is now available in thicknesses of about 8 microns, somewhat thinner than the 13-micron plastic sold in supermarkets. This ultrathin plastic is used as insulation in semiconductors and other electronic devices. Among the common forms of this plastic are Mylar™ and Kapton™. In designing a solar sail for the Halley rendezvous mission, the JPL team considered 25 different plastic films. We studied polyimides, polyethylenes, polyvinyls, polysulfones, and all types of polymers for strength, density, temperature, resistance, manufacturing capabilities, and durability in sunlight and in gamma radiation.

Mylar and Kapton were among the better-known candidates. An 8-micron-thick plastic sail will have a top acceleration of about 0.8 millimeters per second per second—much better than that achievable with 13-micron plastic, but still not sufficient for interplanetary missions, and too low for missions near a planet when the craft will have to fight heavy gravity.

As we noted earlier, we are looking for sail performance in excess of 1 millimeter per second per second. For this we need (taking the payload into account) a sail about 2 microns (0.08 mil) thick. Actually, for very small payloads, such a sail could give an excess of 1 millimeter per second per second acceleration. Since plastic sheets this thin can now be produced, it seems reasonable to consider this thickness of 2 microns for the first *operational* solar sail. (Early solar sail tests may be conducted with somewhat thicker materials, since top performance won't be as important in a test as will be evaluating the actual operation, deployment, and stability of the vehicle. *Test* sails can probably be made of the 8-micron material that is commercially available today.)

Naturally, it will not be possible to build or fabricate a half-mile of sail material on a side in one continuous sheet. However, long strips can probably be fabricated in manageable widths. Sail material might, for example, be made available on a roller, like a giant version of household plastic wrap—say, a half-mile long and perhaps 10 to 25 feet wide.

In the case of the heliogyro sail (a rotating sail made up of narrow strips much like helicopter blades—described later in this chapter), this is ideal; but for large, flat sails, the rolled-up plastic requires us to combine two strips of sail material. This can be done with an ultrathin (perhaps 3-micron thick) adhesive on which strips of sail materials are laid down next to each other, edge to edge. If the solar sail will have to fly through very high temperatures, a coating on the back of the sail material may be necessary so that the sail can radiate its heat off into space and avoid becoming too hot. Thus, the actual fabrication of a solar sail film is fairly complicated, but it can be done in a relatively small laboratory plant. There the sail would be fabricated, the strips would be put together, and then the sail would be rolled up and packaged for a flight into space. A diagram of a magnified sail sheet is shown in Figure 3.1. This is the sail-film design that was planned for the Halley's Comet rendezvous at JPL.

In fabricating sail material, we have to take another possible

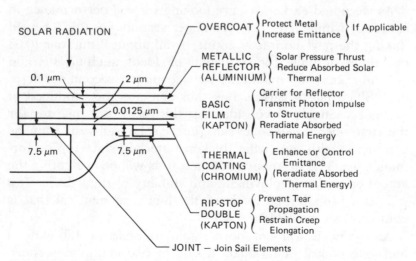

FIGURE 3.1 A greatly enlarged view of the cross-section of a solar sail sheet.

problem into consideration: What happens if a meteorite hits the sail? You might at first think that the meteorite will put just a little hole in the sail, and that as long as the rest of the sail is unaffected, we will be able to proceed with the trip. This is basically true, as long as the hole does not get any bigger, but if the sail becomes shredded, it is not going to do us much good.

We prevent rips from spreading in a solar sail in the same way that the makers of terrestrial sails ensure that a small tear will not become a big one: by building ripstops into the sail material. If you look closely at a sailboat, you will notice that thick seams or tapes are sewn into the mainsail, forming "stripes" across the canvas. These seams, called *ripstops*, are a simple means of preventing rips from spreading. Ripstops can be built into solar sails in the same way, either by doubling the sail material (creating "seams") or by reinforcing the material with tape.

Once we've solved the meteorite problem, though, we have to worry about possible danger from electrostatic charging. After all, as we fly through a region of charged particles, our large metallic sheet will surely pick up a charge and develop currents and short circuits throughout the sail. Because the sail material is thin, voltages could cause arcing and ripping. We deal with this problem by installing an electrical shorting mechanism that runs from the front of the sail to the back, as shown in Figure 3.2. This can be simply inserted at the joints

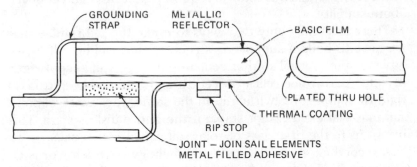

FIGURE 3.2 Solar sail alternatives for electrical shorting to prevent electrostatic charging through the sail.

where the strips of sail material are joined together. With such a shorting mechanism, the maximum amount of voltage differences would be about 5 volts, which would not cause a serious problem in interplanetary flight.

RIGGING THE SAIL

Our next task is to design and construct a vehicle that will permit efficient flight, ensure stability, and enable us to control the sail.

There are many ways to rig a sailboat. Rigging is designed to ensure speed, control, and stability. The solar sail vehicle will be rigged with the same purposes in mind. We will need to steer the craft and keep it steady so it doesn't tumble, tilt, or drift off course. Engineers call these design factors *performance* (measured by the characteristic acceleration or sail density), *rigidization, stabilization,* and *control.* The sail designs that we consider will use different techniques to meet each of these design factors, but they must satisfy each of the factors if they are going to work in space. Theoretically, in the nearly perfect vacuum and weightless conditions of space, we should not have to worry about keeping the sail rigid or stable. But space is not a perfect vacuum, and particles from the solar wind will hit the sail. Also, our steering maneuvers will cause motions (like ripples or flutters) on the sail. These perturbations in space are small, but they do require rigidization and concern about stability.

There are really only two ways to make the sail rigid: either we have to build a supporting structure that will hold it or we have to spin the sail so that centrifugal force will keep it rigid. A rodeo performer relies on centrifugal force to keep a lasso rigid by spinning or twirling it. In the same way, a disk of solar sail material can be kept stable if the entire disk is spun. This brings us to the first class of solar sail vehicles: disk sails.

A second class of solar sail is the heliogyro, which works on the same principle as do helicopter blades. The sails of the heliogyro, like the rotors of a helicopter, stay on the same plane because we keep them spinning rapidly. When they stop spinning, they sag.

If, instead of spinning the sail, we choose to make it rigid by attaching it to a supporting structure, we get to the third class of sail vehicle design, usually called the square sail, although the sails need not be "square" at all; they can have a variety of shapes, including triangles and rectangles.

Figure 3.3 shows these three classes of vehicles, their essential physical characteristics—that is, their method of rigidization, stabilization, and control—and a qualitative description of how they will perform in deep space.

The square sail, or kite-type structure, is held rigid with a supporting structure of wires and booms (Figures 3.4 and 3.5). These booms, as in terrestrial sailing, are called spars. The vehicle is three-axes stabilized—that is, it is stabilized in *roll* (tipping to left or right), *pitch* (plunging up or down), and *yaw*

	RIGIDIZATION	STABILIZATION	CONTROL	PERFORMANCE
● SQUARE SAIL	SPARS	3-AXES	SOLAR PRESSURE VANES	MODERATE
● DISK SAIL	CENTRIFUGAL FORCE	SPIN UNSTABLE BUT CONTROLLABLE	GAS JETS, TORQUE VANES	BEST
● HELIOGYRO	CENTRIFUGAL FORCE	SPIN UNSTABLE BUT CONTROLLABLE	BLADE PITCH	INTERMEDIATE

FIGURE 3.3 Types of solar sail vehicles proposed.

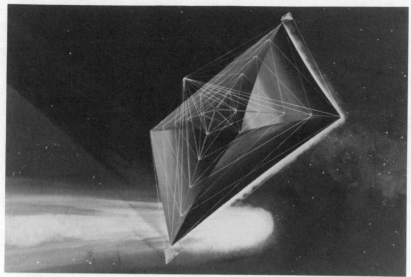

FIGURE 3.4 The square sail, three-axes stabilized and held rigid by the spars, stays, and mast, is the configuration that most resembles the terrestrial sail. On the ends of the booms can be seen relatively small vanes, which are small sails in their own right. They are used to point the spacecraft, that is, to change direction of the sail with respect to the sun. (NASA)

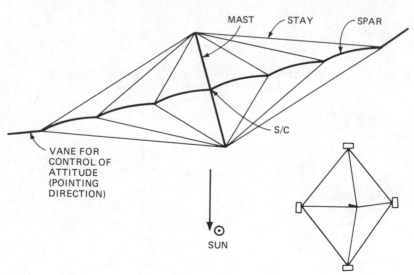

FIGURE 3.5 Square-sail flight configuration.

(veering to one side or the other)—by controlling the forces about these three axes. We achieve this control by using solar-pressure vanes. These vanes are small, independently controlled sails installed at the ends of these axes, something like jibs and mizzens, those small triangular sails in the bow and stern of a larger sailboat. The solar vanes create a force that can move the sail in any one of the three directions. Because a square sail has a structure which, of course, has mass, its performance is moderate relative to the spinning sails, which have less structure. Some of the shapes that have been considered for three-axes stabilized sails are shown in Figure 3.6.

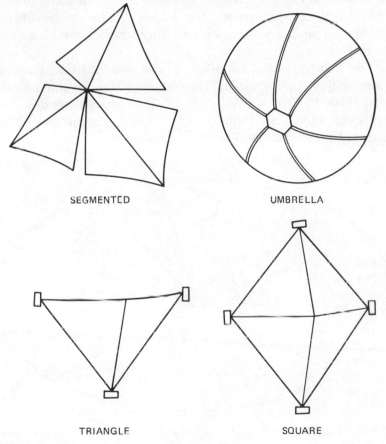

SEGMENTED

UMBRELLA

TRIANGLE

SQUARE

FIGURE 3.6 Types of three-axes stabilized sails.

The first solar sailing concepts were disk sails. They are the simplest, since, in theory, a large sail can be spun and requires no supporting structure at all. Disk sails are kept stiff by centrifugal force. However, it is difficult to control a disk sail by the application of torques from either gas jets or vanes. In fact, the only practical way to control such a sail is to move the center of mass relative to the center of area. This offset creates an imbalance, or torque, that can tip the sail in the desired way. The mechanism, or tracks, for the center of mass offset is, however, complex and heavy. That makes the maneuvering very slow. A possible design for a disk sail is illustrated in Figures 3.7 and 3.8. These also show the elaborate mechanism at the center that allows us to maneuver the payload in such a way that controlling torques can be applied to the vehicle.

The heliogyro is a variation of the disk sail that gives an intermediate performance. This concept was invented in the mid-1960s by Richard MacNeal of MacNeal-Schwendler Corporation and John Hedgepath of Astro Engineering. They adapted their experience with helicopter design in a big way:

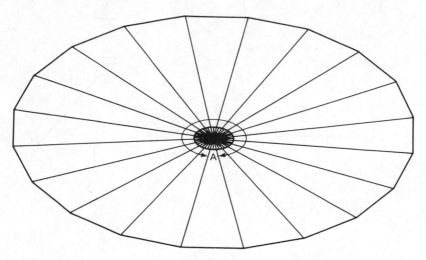

FIGURE 3.7 Spinning-disk sail. The region inside "A" is shown in the detail in Figure 3.8.

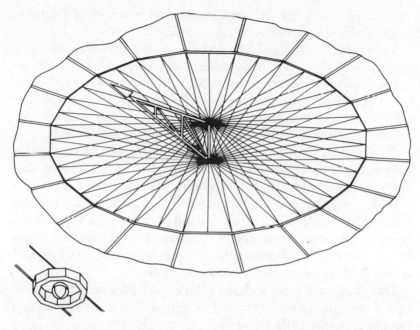

FIGURE 3.8 *Spinning-disk sail center hub. Inner region of the spinning sail. The tracks and boom for moving the payload provide a control mechanism for the vehicle. The center of mass is moved from the center of the vehicle, forcing it to tilt and therefore point in a different direction.*

very long blades of solar sailing material would be constructed and spun like helicopter blades, to keep the sail rigid. The total area of material facing the sun would be the same as that presented by either a disk sail or a square sail, but the heliogyro would have that area configured in long, thin blades. Instead of, say, a square sail 800 yards long and 800 yards wide, the heliogyro sail would have 12 blades, and each blade would be 7,500 yards long and 8 yards wide. Centrifugal force would keep the blades extended without much supporting structure. But the chief advantage to this concept is that it would be easily controllable, unlike the disk sail. Control would be achieved by pitching the blades—that is, twisting the blades' angles from the sun so as to steer the vehicle in different directions (see the Figure 4.4). Thus, the heliogyro is nearly as controllable as the three-axes stabilized sail, and it also has many of the advantages of the structureless spinning-disk sail.

How large does a sail have to be? Recall that the mass per unit area of the sail vehicle determines its acceleration. In an earlier example I mentioned a vehicle with a mass per unit area (or density) of 8 grams per square meter. This vehicle will travel with an acceleration of 1 millimeter per second per second, a reasonable goal for sail materials in the 1980s.

One way to visualize a density of 8 grams per square meter is to multiply that number by 1 million—that is, 8,000 kilograms per square kilometer. In English units, then, we're talking about a vehicle that weighs 17,600 pounds and has an area of about 0.6 mile by 0.6 mile. The thin solar sail sheet, of course, takes up most of this area. A spacecraft weighing 1,000 kilograms might be at the center of this vehicle. The solar sail itself would account for the remaining 7,000 kilograms.

Obviously, we want to have a large sail area to intercept as much sunlight as possible with as lightweight a vehicle as can be built. NASA calls these thin, lightweight structures "gossamer spacecraft"; this is also the designation of the human-propelled machines, developed by Paul McReady, that first flew in the late 1970s. If we can reduce the density to 5 grams per square meter, we get an increase in the acceleration to 1.6 millimeters per second per second—an increase of 60 percent. We wish to make the vehicle large enough so the payload will be a small fraction of the overall mass. To carry 6,000 kilograms (the approximate weight of the *Apollo* command module), the sail would have to be about 2 kilometers on a side in order for the total vehicle to be propelled at the same acceleration of 1 millimeter per second per second. The 2 kilometer-by-2 kilometer sail is equivalent, in the heliogyro design, to a 12-blade sail, each blade 33 kilometers long by 10 meters wide. Think of it: blades 20 miles long, made of material less than one-thousandth of an inch thick, spinning slowly in space! The total vehicle weight would be 36,000 kilograms, 6,000 of which would be the spacecraft.

Conversely, for a test payload of only 100 kilograms (220 pounds), we might need a sail that is only 300 meters on a side in order to get the same acceleration. This will be important in sizing our initial test flights of the solar sail. Thus, as we consider design of the vehicle, we are going to be concerned about large area, thin structure, and as low a mass or density as possible. Multiplying the mass by a factor of ten multiplies the area requirement by a similar factor of ten or the linear size of the sail by a factor of the square root of ten, approximately three. Thus, if you want to move a mass of, say, only 10 kilograms (that is, 22 pounds) around the solar system, the sail (if it is square) will have to be about 100 meters by 100 meters, for an area of 10,000 square meters. The 1,000-by-1,000 meter square sail cited earlier is equivalent to a disk sail about 600 meters in diameter, or a 12-blade heliogyro with blades 8,000 meters long by 10 meters wide.

////4

LAUNCHING
THE
SOLAR
SAILBOATS

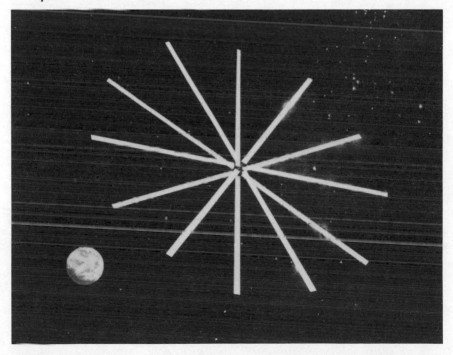

The heliogyro sail as it would look being deployed near Earth. The picture shown depicts the sail halfway deployed. (JPL/NASA)

In addition to worrying about the shape, stability, and size of the solar sail, we have to consider another design problem: Can the sail be carried into space or will it have to be built in space? If the sail has to be carried up in a rocket-propelled ship and then deployed, we will have to give very careful thought to the way we fold, package, unfold, and launch the sail vehicle. This will be a considerable problem in the design of all solar sails. For the disk sail, however, the packaging problem is almost insurmountable, and for that reason we will largely ignore that class of sail from now on. Assuming, then, that we're talking about heliogyros or other nondisk sails, how will we fold them and pack them into canisters that can be carried aloft by rockets? And once we get the packed sails into orbit, how will we unfold them? How will we then rig the solar ship and deploy it in space?

The heliogyro solves this problem nicely by "rolling" the ultrathin sail material blades up and then letting them be unrolled by the spinning motion itself once the spacecraft is in orbit. In fact, it is probable that the sail would be built by rolling the material as it is being fabricated (Figure 4.1) so that it would never be completely unrolled on Earth. This would simplify storage and construction of the sail, and it would probably be necessary, since there would be no way to unroll it in Earth's "1 g" environment without ruining the sail itself.

Thus, in the heliogyro design, the sail is constructed in rolls and unrolled in order to make the blades. Figures 4.2 through 4.4 show a heliogyro sail being deployed. It might take several hours for the blade, which could be more than a mile long, to unroll. As long as the unrolling proceeds symmetrically, the craft will be controllable and stable. This concept could easily be tested out of the shuttle bay during orbital flight, in order to make sure the unrolling proceeds as predicted.

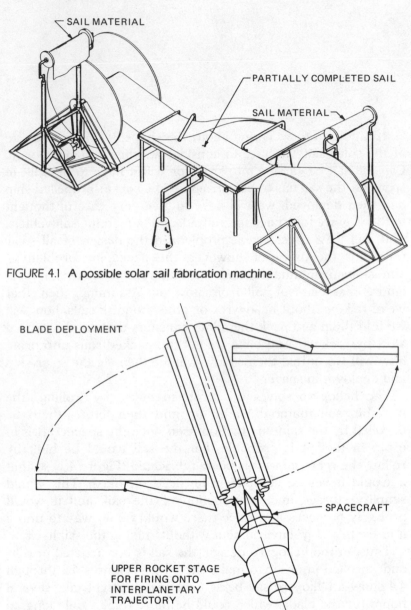

FIGURE 4.1 A possible solar sail fabrication machine.

FIGURE 4.2 Heliogyro deployment sequence. After launch the spacecraft is detached from the shuttle (or other launch vehicle) and the deployment begins. The blades are moved out of the central unit as shown. Details are shown on Figure 4.3.

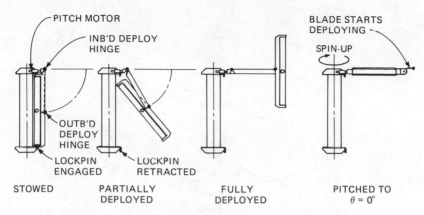

FIGURE 4.3 Heliogyro deployment sequence.

Note that the sail material in the heliogyro does need some sort of thin structure—a thin wire from which the sail hangs and which forms the edges of the heliogyro blade. Those edges carry all the tension for the solar sail and give the sail a great deal more torsional stiffness than it would otherwise have.

Of course, the material used to make these edge-members must be light enough to permit the solar sail to fly. The Jet Propulsion Laboratory design team used graphite polyimide tapes as edge-members. Many tapes, each one only 2 by 0.2 millimeters in size, were used in order to provide redundancy in case one of them snapped, for example, after being hit by a meteoroid.

Terrestrial sailors will need no explanation for the use of battens across the blade at approximately 100-meter intervals. Battens are strips of wood or light plastic inserted in the sail to keep the sail material stiffer. In spacecraft we'll use plastic or metal battens. Between the battens, the edge-members hang like catenaries—curves like the ones shown in Figure 4.5. Thus, the heliogyro blade is not simply a sheet of material several miles long. It is still relatively simple, though. The battens and tapes support the sail as it unrolls and as it is pitched.

The terrestrial sailor will also immediately understand the method of deploying the square sail, which is much like hauling up a mainsail. It is done with masts, spars, winches, and guy

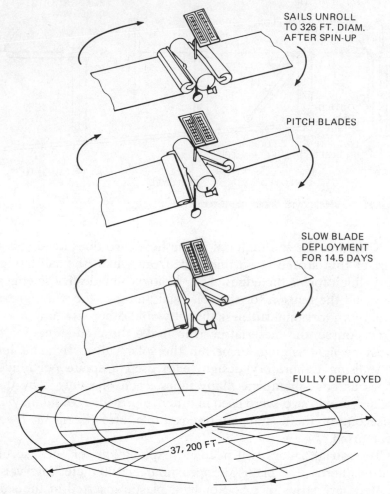

SAILS UNROLL
TO 326 FT. DIAM.
AFTER SPIN-UP

PITCH BLADES

SLOW BLADE
DEPLOYMENT
FOR 14.5 DAYS

FULLY DEPLOYED

—37,200 FT—

FIGURE 4.4 Heliogyro deployment sequence unrolling the sail (blades). Diagram showing deployment and pitching of two of the heliogyro blades. The pitching of the blades changes the angle of the sail with respect to the sun.

wires. Figures 4.6 to 4.9 show the process of square-sail deployment. Simply stated, the masts and spars are first deployed, probably by extending a telescopelike device or by unrolling prestressed metal spars until they reach their full length of several hundred meters. As this prestressed metal unrolls like

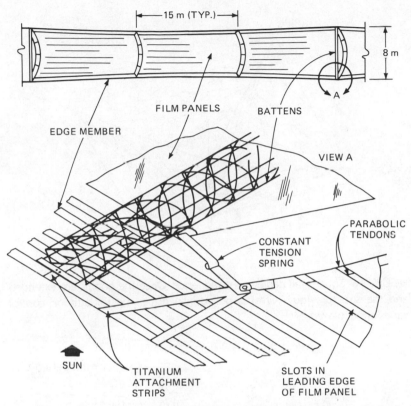

FIGURE 4.5 Baseline design of a heliogyro sail blade. The blade of a heliogyro is more than just a strip of material. Graphite polyimide wires provide edge-members on which to "hang" the sail. Battens provide stiffness and, built in a slightly curved shape, they prevent wrinkling of the material.

a roll of toilet paper, the strip begins to curl so that it forms a long tube that will serve as a boom—the swinging "pole" that is attached to the foot of a terrestrial sail. Guy wires attached to both ends of this boom will then begin to unfurl the sail, which will have been folded into a canister. The wires will pull the sail straight out by its corners and hang it on the spars. This work will probably be done by robots, perhaps with human crews monitoring the deployment and standing by to perform any necessary "unsnagging." The guy wires provide added stability to the overall structure and keep it from flapping around and

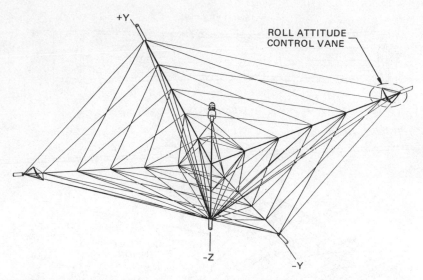

FIGURE 4.6 Square-sail deployment sequence. Once the booms are extended and the structure (guy wires) is pulled into place, the roll-attitude control vanes are deployed.

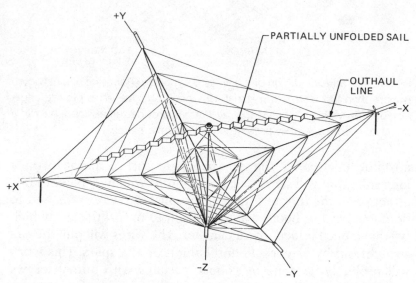

FIGURE 4.7 Square-sail deployment sequence. Sail x-axis outhaul: With the structure in place, the sail itself can be unfurled first by pulling in one direction (x) and then perpendicular (y).

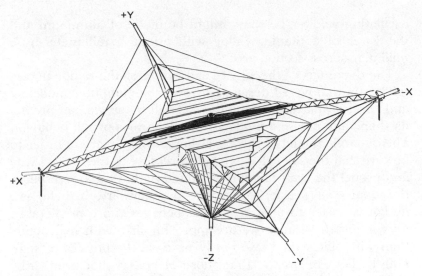

FIGURE 4.8 Square-sail deployment sequence. Sail *y*-axis outhaul.

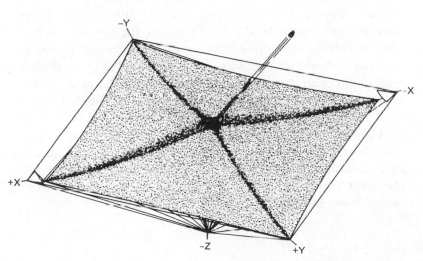

FIGURE 4.9 Square-sail deployment sequence. Final cruise configuration.

oscillating wildly. The spars might be made of aluminum and the wires of a titanium alloy with about ½ millimeter-by-½ millimeter cross-section area.

The dynamics of the square sail is one of the major uncertainties in the overall design. However, by computer modeling and careful analysis of how the sail will hang, we can predict its shape and devise a model of the dynamics of sail behavior. The dynamic model of the sail's shape is necessary in order to be sure that the sail will catch light pressure in the correct way and propel the craft along a predicted path. In many respects, the square sail is easier to visualize, but in reality, the heliogyro design is easier to analyze, and it behaves more predictably. This is partly because we are more familiar with helicopters than with "kites" and also partly because the physics of spinning bodies are simpler than those of bodies that need to be artificially stabilized.

The JPL team finally selected the heliogyro design for these reasons. However, most engineers wish to do further work on the square-sail design, since its linear dimensions are smaller and the square sail may be more controllable once we understand the analysis.

If we don't have to worry about packaging and deploying the sail—that is, if we can construct it in space, from a space platform or other orbital manufacturing facility—the job of a solar sail design becomes even easier. In zero gravity, a crew working in a vacuum could build and assemble large sheets of material in orbit without any deployment mechanisms. Robotic building may be an easier alternative. To do this will require orbital manufacturing capability, however, so we will have to wait until this technology has been developed. Nevertheless, it does become important as we consider the future of solar sailing, including the use of a solar sail vehicle as an interplanetary shuttle or ferry. Then the sail will be used repeatedly, to and from Earth orbit and on into deep space.

5

"SAILBOATS" VERSUS "POWERBOATS"

An artist's concept of a nuclear-electric propelled spacecraft.

As I pointed out earlier, the development of the solar sail in NASA became caught up in a competition with ion-drive or electric propulsion. Continuous low thrust, whether it comes from a sail catching sunlight or from an engine powering the spacecraft, provides performance advantages over ordinary chemical propulsion on long interplanetary trajectories—but the advantages must be balanced against weight requirements for the propulsion system and against the complexity of spacecraft design.

Those of us in the solar sail business would like to argue in favor of the sail on purely aesthetic grounds. It requires no propellant, uses free sunlight, and allows for beautiful gossamer designs. Like terrestrial sailors, we sneer at electric-propulsion craft, which require poisonous propellants, rely on advanced and complex engines, and must carry along an inefficient power-conversion unit. But we need more than aesthetic arguments to make hard-headed choices for interplanetary travel. We must know more.

An electric-propulsion system has three basic parts: (1) the source of electricity and (2) the power conditioner that converts electric power into just the right current and voltage for ionizing the gas in (3) the ion thruster. See an ion rocket engine in Figure 5.1. The electricity can come from solar energy through solar cells. The power processor is a high-power electronic system. The ion thruster operates by sending a gas (propellant) into a chamber where the atoms of the gas are electrically ionized. The ions are then channeled through a screen and an accelerator electrode to form an ion beam. The beam is neutralized by electrons coming from the conditioned electric power. The electric energy increases the propellant's exhaust velocity and hence its thrust. The propellant must be a

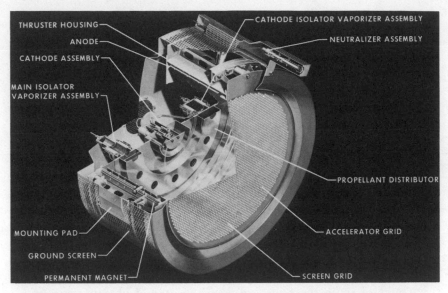

FIGURE 5.1 An ion rocket engine: Liquid mercury propellant enters reaction chamber through main vaporizer assembly; vaporizer assembly heats liquid producing mercury vapor; vapor is distributed to reaction chamber through propellant distribution manifold; and cathode assembly emits electrons that impact and ionize mercury vapor atoms. Resulting ions pass through screen grid into acceleration electric field between screen and accelerator grid.

heavy metal that readily gives up its electron in its gaseous state—the ionized plasma. Mercury is the usual propellant. Unfortunately, it is a dangerous material both on Earth and in space. Current attention is being paid to xenon and krypton, which are inert and nontoxic (despite Krypton*ite*'s bad effect on Superman). The weight of the total propulsion system is the sum of the propellant and the power plant (engines, power conditioners, and ionization chamber). Since the power-plant weight requirement goes up rapidly with very high exhaust velocities, the electric power system is inherently limited.

A solar sail has infinite exhaust velocity, since it uses no propellant. (It does not, of course, have infinite thrust). Thus, we feel that the sail vehicle is less limited for the very large jobs of solar system exploration.

Ion thrusters were tested in space in the 1960s and early 1970s. The lifetime of thrusters is a concern and so their qualification for flight use will take a long time. We do not know how long they will work in space, and until we do we can't plan on their use in interplanetary flight. Like solar sailing, electric propulsion (with much smaller systems) can be used to control the attitude of spacecraft, rather than as primary propulsion. Several Earth-orbiting spacecraft designs include electric-propulsion attitude-control systems.

The key parameters of an electric-propulsion system are the electric power of the source—either the solar-cell array or the nuclear reactor; the weight of the power plant, mentioned earlier; and the weight of the propellant. The performance specification of electric-propulsion systems (analogous to characteristic acceleration for the sail) is mass per unit of power—the lower the better.

For the Halley's Comet rendezvous mission the ion-drive system requirements for these parameters were: 70 kilowatts at Earth, less as the spacecraft traveled away from the sun; 1,700 kilograms for the power plant; and 2,400 kilograms for the propellant (see Figure 5.2). The solar array required a concentrator—a parabolic mirror that focused more sunlight on the cell array. The parabolic mirror was itself a thin reflector, somewhat like a much smaller version of the sail, fitting around the solar panels. Approximately 1.2 million solar cells were required to be arrayed over nearly 500 square meters. The array was to be spread over two wings, each 74 by 3.3 meters. Each parabolic concentrator was 74 by 15 meters. The entire array was to have been deployed in a series of unfolding maneuvers after the spacecraft was launched and deployed from the shuttle.

This system probably could have been built. But it never would be. It would meet the Halley mission requirements, but with enormous complexity. Solar-electric propulsion has been under development in NASA since 1959—during the whole history of the space age. The Halley's Comet system was far different from the solar-electric systems that were earlier considered for interplanetary flight. The 70-kilowatt power was five

FIGURE 5.2 *The ion-drive system designed for a Halley's Comet rendezvous.*

to ten times larger than those systems. The present proposed system for a comet rendezvous has 10-kilowatt power, a 70-kilogram power plant, and carries 880 kilograms of propellant.

Solar-electric propulsion won the competition over the sail because it was considered more "technologically mature." But within weeks the victory was deemed Pyrrhic. NASA dropped the Halley rendezvous mission. Within a few months, it also dropped all Halley mission ideas and all low-thrust propulsion development. Unfortunately, solar-electric propulsion may now be an idea whose time has come and gone.

So where are we now? Little solar-electric propulsion development for interplanetary missions is going on in the United States now. The Germans are studying it for a comet rendezvous mission of the European Space Agency. Since the recent U.S. launch failures, there has been a new interest in the use of solar propulsion for the U.S. comet rendezvous mission also. But this seems like a repeat of earlier U.S. studies. Technological advances in solar-cell efficiency, propellants, or power conditioning may provide a breakthrough, but in my judgment, the alternative of gravity-assist cleverness is progressing faster. Worse yet, the pace of interplanetary exploration to deep space (beyond Mars and Venus) is slowing so much as to provide little reason to develop a good solar-electric system, and this is just as true for solar sailing. I think by the time we need solar-electric propulsion it will be outdated and will have been replaced by nuclear-electric propulsion. At that point, the low-thrust choice will be between NEP and solar sailing.

Nuclear-electric propulsion replaces the solar-array power source with a nuclear-reactor source. (See artist's rendering on page 57.) A "small" reactor is 100 (or even hundreds of) kilowatts and isn't, of course, dependent on distance from the sun. High power is of interest to many space applications of the future, both civilian and military; hence, its availability from a nuclear source for both propellant procedure and power is an advantage. But nuclear power has both real and imagined risks, and one of these might delay its application.

The explosion of the space shuttle and the accident at Chernobyl, both in 1986, heightened concern about the launching of nuclear power sources into space. The *Galileo* mission to Jupiter uses radioisotope thermoelectric generators for power—a nuclear source that supplies power from decay of radioactive atoms, not from the nuclear reaction. These devices are low in power, small, easily packaged, and incapable of explosion. Yet even they involve controversy and risk. We need nuclear power to explore the outer solar system and probably even to work extensively on Mars and the moon. I am confident we will find ways to minimize the risks during launch and in space. But it won't be cheap.

The high-power levels of nuclear-electric propulsion offer exciting possibilities for the future. These possibilities are competitive with solar sailing in the inner solar system and advantageous for missions beyond the asteroid belt. Their by-product of high power makes them even more useful. But the solar sail still has the advantages of cleanliness (no nuclear sources), no fuel or consumables, and probably a low cost. It will be a long time before we can use the sail in the inner solar system, however, so worrying about its performance disadvantages in the outer solar system may be premature. Furthermore, as we will see, the sail (with lasers and gravity assist) may be the best means of interstellar flight. If we can use sails to go to the stars we certainly will be able to use them to transport material and instruments to the outer solar system—for orbiters, satellite landers, and other spacecraft concepts.

//// 6

NAVIGATING
TO
THE
PLANETS

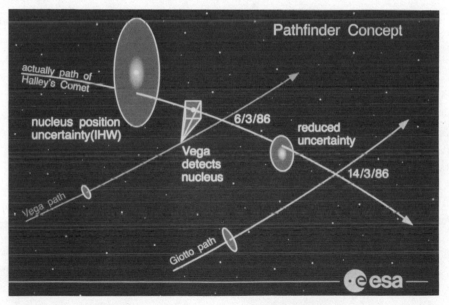

One of the outstanding successes of the international effort to explore Halley's Comet was the cooperative Pathfinder Navigation project of the USSR, Europe, and the U.S. Soviet comet pictures and American and Euopean spacecraft tracking were used to retarget *Giotto* closer to the comet. (European Space Agency)

Navigation consists of three basic elements: You must know where you are, be able to predict where you are going, and be prepared to make any necessary corrections so that you will end up where you want to be. These skills are fundamental to all navigation, whether on a ship, in an airplane, across a land area, or between the planets. Over the centuries, various navigation tools and techniques have been devised, but the fundamentals have never changed.

In order to know where you are, you have to do two things: First, you have to make some measurements; and second you must translate those measurements into agreed-upon coordinates. If you are using celestial navigation, for example, you will look at the sky and measure the positions of the stars in azimuth and elevation. Azimuth is the angular distance from north along the horizon. Ninety degrees azimuth is east. Elevation is the angular distance up from the horizon. (*Altitude* is the correct term; *elevation* is more often used, however.) Ninety degrees elevation is directly overhead. Then using tables, equations, computers, slide rules, or other tools, you translate these positions into latitude and longitude. This is possible because there is a mathematical relationship between latitude and longitude and those of star azimuth and elevation.

You can take these measurements with instruments such as a sextant, which measures angles, and a chronometer (a watch), which measures time. Time relates azimuth to longitude. To determine your latitude and longitude on the surface of the Earth, you need two measurements. If you are trying to figure out your position in space, you need at least three measurements.

In actual space travel, however, navigators usually try to figure out a minimum of six coordinates: three of position and three of velocity. Remember that position and velocity are independent of each other. You and I can be in the same place, traveling at different speeds, or we can be traveling at the same speed in two different places. So we need a minimum of six measurements to unambiguously know where we are and where we are going. If we are fortunate enough to get more than six measurements—that is, redundant measurements—we can use the extra ones to get a more precise estimate of the coordinates that we are trying to figure out.

All measurements, of course, contain errors due to the impreciseness of the equipment or to an observer error. These measurement errors translate into coordinate errors. You can use redundant measurements to reduce the errors in your calculation of position and velocity in a commonsense way: simply take the measurements over and over again, counting on the fact that the random errors in the measurements will average out.

In modern navigation, we call this process *filtering*. We make many "noisy" measurements (measurements with errors), put them through a "filter," and combine them all into the optimal (least erroneous) determination of position and velocity—or any other coordinates we are seeking to determine.

Filtering is an electrical engineering term. The filters in our radios and television sets bring out the characteristics of the signal to the best advantage in an environment that has static or noise. Navigational filters serve the same purpose, except that the process need not be electrical or electronic; it can be analytical or computational in that it takes the data and extracts the maximum amount of information from it while rejecting or minimizing the noise (errors).

Suppose, for example, that a spacecraft is traveling on an elliptical path between Earth and Mars. We may make many measurements in order to deduce the position and velocity of the spaceship. The results of those measurements should follow a smooth path of position and velocity that obeys Kepler's laws of ellipses. This is the signal. The deviations from that ellipse could be due to noise in the measurements. We need to filter out that noise in order to deduce the spacecraft's position and velocity. We then use those measurements to estimate the parameters of the ellipse, and from the measurements we deduce statistics that tell us how precise our calculations are.

I mentioned that one type of measurement common to celestial navigation is to measure star angles and time. For centuries this was the mainstay of shipboard navigation, and naturally, when we first began flying, this was the most common measure of airborne navigation, other than looking out the window at landmarks.

With the advent of radar, two very important means of measurement were introduced: ranging and Doppler. Ranging is a measurement of the time it takes for a radar signal to leave the craft, bounce off a distant object, and return to the sender. That time interval can be precisely converted into a measurement of distance, or range, because we know that a radar signal travels exactly at the speed of light; in fact, all electromagnetic radiation travels at the speed of light. By multiplying the speed of light by the time interval, you can determine the range. The Doppler effect was deduced in 1842 by Christian Doppler, a German physicist. He figured out that the frequency of a signal is affected by the speed of the object that is emitting the signal. Frequency is the distance between successive waves of the emitted signal. If the object emitting the signal is moving away from you, then the distance between waves increases (and their frequency decreases). If it is moving toward you, the signal starts catching up to itself and the wavelength distance decreases (and the frequency increases). This frequency shift lets *you* determine the relative speed between you and the emitter.

These frequency changes are called *redshifts* for signals lower in frequency (or moving toward the red side of the light spectrum), and *blueshifts* for those higher in frequency (or moving toward the blue side of the spectrum). We observe distant stars and galaxies by measuring the spectra or frequency of the light they emit. The atoms and molecules making up the stars and galaxies vibrate at distinctive frequencies, and the presence of those atoms and molecules is revealed as lines recorded on the photographic plate of a spectrograph. When the star or galaxy is moving away from us, the frequencies of these lines appear redshifted; if the object is moving toward us, they are blueshifted.

If an aircraft or spacecraft moves away from us, its frequency will also be shifted, and we can measure this Doppler effect if we know what frequency to expect and compare it to the frequency we measure. From the frequency shift we can deduce the velocity. Thus, we get two types of measurements from the radio navigation of spacecraft in deep space.

If we make these measurements repeatedly, we can deduce our position and velocity, and the redundant measurements allow us to reduce the errors to minimal values. The accuracy of spacecraft navigation is extraordinary. Range over millions of miles can be measured to the accuracy of nanoseconds—that is, one part in a billion seconds. Similarly, the velocity, or Doppler shift, can be measured to accuracies also better than one part in one billion, and filtering reduces the error in position and velocity to even less. This kind of precision is necessary in deep space, for small errors can build up into huge "misses," and the job of correcting, for even small errors, can turn out to be expensive.

This is only the first of the three principal elements of navigation—measuring where we are. Our next job is to predict where we are going. To do this we use equations of motion—precise mathematical descriptions that tell us how flight occurs.

The basic way we move about the solar system is, of course, dictated by the law of gravitation, which was first enunciated by Sir Isaac Newton. That law has been refined by the general theory of relativity developed by Albert Einstein. We need that refinement in order to make accurate predictions of motion, although for much of our everyday experience, Newton's description of gravity suffices. It is also important not to limit our calculation to that elliptical motion of the spacecraft about the sun; we also have to take into account all of the perturbations to this motion from small effects that we might otherwise not think important.

For example, we need to understand the effect of other

planets besides those we are near. Sometimes we must even include the effect of moons or asteroids, and we have to account for the solar wind (remember that this is not the same as solar pressure). If we are near a planet, we might also have to take into account the very small drag from its upper atmosphere. If it is a nonspherical planet, we need to know the effect of its bumpiness and lumpiness on its gravitational field. We then put all of these data into equations of motion, and we use very fast and accurate computers to solve those equations. In the end we come up with a precise prediction of our position and velocity at a given time, or a precise prediction of the time we will reach a specific place.

Thus, for example, on our trip to Mars, we first measure our position and velocity. Then we deduce where we will end up in relation to Mars. If this is not where we want precisely to be, we need to go to the third step of navigation: guidance—making the necessary maneuver to get us back on the flight path so we end up where we want to be.

We can do this by using the *delta vee* (ΔV), which is shorthand for change in velocity. This is the change we make—perhaps by firing a rocket—in order to aim our ship in the direction we wish to go. We try to end up on a course that will take us where we want to go.

Thus, we have proceeded through four stages: measurement, orbit determination, orbit prediction, and guidance—or steering. The analogies with terrestrial sailing are obvious. On Earth we deduce our position, figure out where our present course will take us, compare that to our desired course, and compute a path to get back.

Our job is complicated, however, by the fact that the sail vehicle is not just a ballistic spacecraft flying through space under the influence of gravitation; it is a low-thrust spacecraft principally propelled by solar pressure. Thus, our equations of motion are somewhat more difficult and our velocity corrections, or steering laws, are more complicated. Our equations of motion must take into account the physics of the solar photons that hit the spacecraft and move it through space. We steer our ship by moving the solar sail so that the velocity change is built up slowly. For a conventional spacecraft, the velocity change would be instantaneous. In a solar sail vehicle, however, it is built up over time by the small acceleration we use to get where we want to go.

Thus, navigation becomes a continuous process on the solar sail spacecraft, and we must steer continuously, not just occasionally, as in a ballistic spacecraft. Like terrestrial sailors, we will take position fixes intermittently, but a good helmsman will navigate constantly in search of the best way to get us to our destination on time.

IIIIII 7

THE

INTERPLANETARY

SHUTTLE

The interplanetary shuttle, dropping a payload at one planet and continuing on regularly to and from the planets. Since it uses no fuel and since the payload masses are so large, the sail is ideally suited for such operation.

At the Jet Propulsion Laboratory in 1976, Jerome Wright first developed the idea of the interplanetary shuttle. He wrote:

> This vehicle is envisioned as being a reusable solar sail which would . . . deliver spacecraft to various planets, asteroids, short period comets, or solar orbit. The sail may carry multiple payloads on a single mission and, after completing all of its deliveries, would return to Earth, then spiral down to a low Earth orbit for its next mission. While the sail is in this parking orbit, it may undergo any necessary repairs or refurbishment prior to its mission. If a solar sail is developed for use with the Halley's Comet mission, it may be feasible to design the sail module in such a manner that it can readily be adapted to a reusable configuration.
>
> The payload capability of a solar sail to Mercury is quite high by current standards. With a flight time on the order of 900 days, a sail the size of one used for a rendezvous with Halley's Comet would be capable of carrying payloads of approximately 10 tons or greater. The sail would enable the return of a sample from Mercury, and if used at Mars, could probably provide for the return of a sample significantly greater than what could be achieved by purely ballistic means. A Mars lander of 5 or 6 tons might be delivered by a sail of the design developed for a Halley rendezvous.

To see how the solar sail will perform on missions throughout the solar system, we must measure how large a payload it can deliver in various places. Let's start on Earth.

The space shuttle can deliver about 25 tons to low Earth orbit; but it cannot go higher or farther than that. To extend its range, the shuttle would have to carry a rocket or added propulsion device as part of the 25 ton payloads. The amount that this rocket must weigh depends principally on the kind of fuel and the delta vee—that is, the velocity change we want to apply. The velocity change, in turn, depends on how far we want to go.

It would be nice if we could dispense with the rocket, immediately deploy the sail, and begin using it to propel the craft to higher orbit or away from Earth. But we can't. At shuttle altitudes (about 160 miles), a tiny bit of air will cause the gossamer sail to be dragged into the atmosphere. Even if this weren't the case, it would still be difficult to sail out of low orbit. We have to keep orienting the sail toward the sun while flying in a 90-minute orbit. It is difficult to move the sail that fast, and we get so little push out of it on each orbit that it takes a very long time to get away. Although a trip from high Earth orbit to Venus might take only 90 to 100 days by sail, the trip from low to high orbit around Earth would take almost one year. It would be like trying to maneuver a 90-foot sailboat out of a huge marina under sail—a slow, tricky job that could take many hours. The wise crew will leave the marina under power and hoist sail only when the craft is out in the open sea. Similarly, the solar sail ship isn't meant to be used in busy "harbors"; it is most efficient in open space.

Thus, we will use rocket propulsion to get from low Earth orbit, near or on an escape trajectory. This trajectory is not a closed ellipse around Earth; it is a path that leads away from Earth indefinitely. In reality, it eventually becomes a closed orbit about the sun, but it allows us to escape Earth's orbit. Once we get on our interplanetary orbit, we can use the sail for acceleration.

Figure 7.1 shows some of the payload capabilities of a sail like the one designed for the Halley's Comet rendezvous mission. Various factors such as sail size, sail thickness, payload mass, flight time, and launch velocity all affect mission design in complicated ways. In this chapter, we assume that the values stated are feasible and realistic. This is a square sail about a half-mile on a side, or a 12-blade heliogyro with 3.5-mile–long blades. The sail material is 0.1 mil thick. The characteristic acceleration is about 1 millimeter per second per second. Naturally, the accelerations are higher nearer the sun, and therefore the performance to Mercury and Venus would be better than that to Mars and Jupiter. The sail could deliver 8,300 kilograms (18,300 pounds) to Mercury in 600 days, 17,000

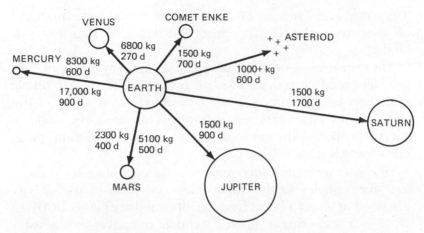

FIGURE 7.1 Capability of the sail vehicle for interplanetary exploration (640,000 square-meter sail, 2.5 microns thickness). Payload mass in kilograms; trip time in days.

kilograms in 900 days, and an amazing 42,000 kilograms (almost 100,000 pounds) in 1,500 days (4.1 years). Even allowing 3,000 kilograms for orbiter propulsion, the sail could accommodate a heavy orbiter and several landers. It could deliver 6,800 kilograms to Venus in 270 days. This means that the solar sail would enable us to design and carry out a Venus sample return in less than 3 years.

But the sail would also provide adequate and exceptional performance to Mars and even beyond. It could deliver 6,000 kilograms to Mars, permitting sample return missions, since it could travel back and forth between planets without using any propellant. We could use the sail to obtain a sample return from the Comet Encke in four to five years. For this mission we would design a 900-kilogram main spacecraft and a 400-kilogram probe, or daughter spacecraft, to stay at the comet. On a comet mission, unlike a trip to Mars or Venus, we would not need the extra landing and takeoff vehicles.

A solar sail, furthermore, could deliver 1,500 kilograms to Jupiter in 2.7 years and to Saturn in 4.8 years. On this mission the spacecraft could carry an orbiter and at least one atmospheric probe. At Saturn, for example, we could concentrate on

Titan, one of the moons. Perhaps we could do a radar mapping of Titan from the Saturn orbiter and from a Titan probe or lander.

These are envisioned as unmanned one-way missions, but the sail could be programmed to return to Earth. A multiple main belt asteroid survey could be carried out at a cost of trip time—about two years per asteroid. This could be done in several probes, or the sail could carry one 1,000-kilogram spacecraft for this main belt survey.

One way to further increase the sail's capability for delivering large masses to the outer solar system is to use gravity assist—that is, get a boost from an intermediate planet by flying close to it and using its gravity to boost ourselves into a faster trajectory. This principle is a consequence of Newton's laws of motion, but it was not analyzed until nineteenth-century mathematicians studied the orbits of comets. When a comet passes close to one of the giant planets, its trajectory is permanently altered by the gravity of the larger body; in other words, it receives a gravity assist. The short-period comets—those that regularly return to the inner solar system—have been diverted into these short orbits by gravity assist.

With the advent of space flight in the 1950s and 1960s, celestial mechanicians began to look more closely at the concept of gravity assist. Mike Minovitch and Gary Flandro at JPL, D. F. Lawden in England, K. A. Ehricke at General Dynamics, Walt Hollister and Richard Battin at MIT, and G. A. Crocco in Italy were pioneers in the field. They discovered that gravity assist was more than a perturbation to a trajectory that must be considered in calculations; these men saw that the assist also provides a means of controlling or shaping a trajectory.

The designers of the *Mariner 10* mission were the first to use gravity assist to reach a planet. The spacecraft used gravity assist from Venus to reach Mercury. It then used another gravity assist from Mercury to return to Mercury on another orbit. *Pioneer 11* and the *Voyagers* used a gravity assist at Jupiter to reach Saturn; *Voyager 2* encountered Uranus and will encounter Neptune with help from the gravity of Jupiter and Saturn. *Galileo* will use gravity assist to help it into orbit around

Jupiter and to tour among that planet's moons. The Soviet Venus–Halley mission (*VEGA—VE* for Venus and *GA* for the Russian spelling of Halley) used it to redirect a Venus mission on to Halley's Comet. How does gravity assist work? When a spacecraft on a trajectory about the sun (flying from one solar system body to another) passes close by a planet or moon, its trajectory is bent by the gravity of that body. If it flies ahead of the planet (in the direction the planet is moving), the spacecraft's trajectory is bent back and it loses velocity. If it flies behind the planet, it gains velocity.

This sounds like getting something for nothing—a violation of the law of conservation of energy. How can the spacecraft pick up speed without doing any work? In its motion relative to the planet, the spacecraft comes in and leaves with the same speed and the same energy relative to the *planet*, so there is no violation. But velocity is speed *and* direction, so when the spacecraft changes direction, its velocity relative to the sun is increased or decreased. The energy of the spacecraft and its trajectory about the *sun* are both changed. It has received a gravity assist.

Since the total energy of bodies in orbit around the sun must be conserved, if a spacecraft gains energy relative to the sun by gravity assist from a planet, then the planet must lose energy. Jupiter and Saturn lost some orbital energy when *Voyager* flew by them! Fortunately, the loss was in the ratio of the mass of the spacecraft to the mass of the planet, or about 0.00000000000000000000004 percent. We leave it to future generations to worry about the effects of ten trillion billion spacecraft flying by a planet or moon.

If we are fortunate enough to have Jupiter in the right position, as it was for the *Voyager* flights, we can use it to boost us to Saturn, Uranus, Neptune, and Pluto. This will reduce our flight time and improve our mass-carrying capability. But this only happens once every two decades for Saturn. We can also use Venus and Earth for gravity assist by looping around the inner solar system before going on the outer planet trajectory. Although this adds to trip time, it does improve performance. These Venus–Earth gravity assist (VEGA) trajectories were discovered by Gerry Hollenback at Martin-Marietta Corporation in the mid-1970s. The gravity assist by the Galilean satellites of a Jupiter orbiter is of the same type, except that it is in the mini-solar system of Jupiter rather than that of the sun. The VEGA-type trajectories are very much suited to the interplanetary shuttle as illustrated in Figure 7.2. The sail vehicle flies on continuous trajectories to the inner solar system and Mars, delivering and receiving payloads and occasionally sending spacecraft to the outer planets.

Because of the greater force of the solar pressure near the sun, we can gain performance advantages by directing the sail vehicle first inward, then outward. A trajectory inward can provide sufficient velocity to enable long looping ellipses to the outer planets. Combining with gravity assist while passing by

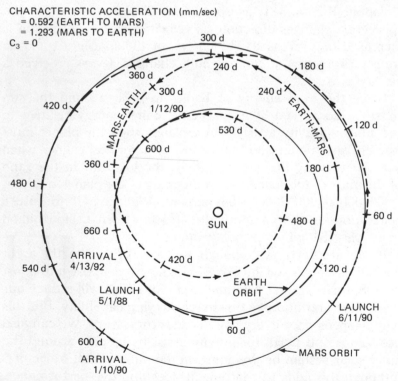

FIGURE 7.2 An example of a 1988 solar sail Mars sample return mission.

planets on these ellipses can improve performance more. As we will see, it is best to go inward first, even if we're trying to go outward.

These examples are titillating in that they show how enormous missions are possible with the sail. This is important, for the mission requirements for future exploration keep increasing as we learn more and want to do more. This drives the cost of individual missions up, and so fewer and fewer missions are approved. But if the solar system is ever to be a place of human activity and real exploration, we must find a way to conduct economical, regular, and large missions. This is the promise of the solar sail.

And so we imagine that space is an ocean and that the outer planets are its shores. We will explore this ocean, and the sail will take us to those distant shores. An Earth orbit facility, perhaps a space station, can be our shipyard. At first it will be a place from which to deploy the sails we bring up from Earth, but later we will construct the sails and rig the spacecraft right there at the space station.

The sail fabrication might first include deposition of aluminum on plastic as I described earlier, but our goal will be to get rid of the plastic substrate and use only a metallic film for the sail. The shipyard will probably need human workers, but most of the construction will be done by robots.

Spacecraft will also return to this port, and the space station will handle payloads on their way to or away from Earth. The sail will be repaired at the space station if necessary—patches will cover micrometeroid holes, for example—and then be used again for other solar system missions.

The solar system fleet will include spacecraft of several sizes, designed for various purposes. There will be supply ships for the inner planets and asteroids, faster craft for missions to the outer planets. The sail spacecraft might carry other exploration vehicles—landers, atmospheric balloons, Mars airplanes, rovers to work on planetary surfaces, *in situ* lander stations and vehicles that can collect samples at the planets, satellites, asteroids, or comets and return them to us.

Let's consider Mars. Although it is not our nearest neighbor, it is the most likely place in the solar system for extended human activity. It offers a wide range of exploration objectives and incentives, ranging from geology, climatology, meteorology, and geophysics to the potential for human colonization. Mars also permits us to consider various exploration activities—landers, rovers, airplanes, surface facilities, *in situ* propellant, and power production for both machines and humans. Since Mars is outbound from Earth, it also offers a good opportunity to use the solar sail. If there is to be a role for an interplanetary shuttle, Mars missions will be a primary design requirement and the sail will have to be specified to meet such requirements. Venus and Mercury missions will, of course, be much easier. Finally, although lunar, asteroid, and artificial satellite bases are conceivable, it is likely that they will at best be outposts and primarily robotic stations for science or material supply. Mars is our goal if we are to think about humans permanently or self-sufficiently operating elsewhere than planet Earth.

Round-trip missions to Mars (or anywhere else) involve three main stages: the Earth-to-Mars transfer, the stay time at Mars, the Mars-to-Earth transfer. Here are the breakdowns of those three stages:

1. Earth-to-Mars Transfer:
   - Leave Earth orbit—spiral out with the sail or accelerate with rocket away from Earth; then deploy the sail
   - Earth-to-Mars trajectory—flight time depends on acceleration, hence on payload mass; can delay arrival to permit less stay time at Mars
   - Mars arrival—go into orbit first to utilize sail vehicle for return trip; spiral into orbit
2. Mars Stay:
   - Entry
   - Descent
   - Landing
   - Surface exploration—sampling, limited mobility or extended rover capability
   - Sample collection

- Liftoff
- Ascent—enter into Mars orbit and rendezvous with return vehicle (the sail)
3. Mars-to-Earth transfer:
    - Await alignment of Mars–Earth geometry if not yet proper or begin on slower return trajectory instead of waiting; spiraling out from Mars orbit will take months unless ascent is to very high orbit for rendezvous
    - Mars-to-Earth trajectory—flight time depends on acceleration
    - Earth capture—capsule can be separated for direct entry into Earth or carried with sail into orbit; spiral into orbit around Earth, rendezvous with station (platform, shipyard), in "port"

Interplanetary missions are critically dependent on the geometry of the planets if we are using ballistic spacecraft, but for low-thrust missions we can continuously modify the orbit and optimize the trajectory for rendezvous at the target. Analysis of interplanetary missions to take this into account has not yet been done extensively, but it will be if the interplanetary shuttle is to become a reality.

In general, with a sail characteristic acceleration of about 1 millimeter per second per second, flight time to and from Mars will be less than 400 days, but we cannot reduce the trip time much below this estimate. As we noted earlier, planetary geometry is the more important factor shaping interplanetary trajectories. A four-year round-trip mission example (1988) is shown in Figure 7.2. The outbound characteristic acceleration, in this example, is only 0.592 millimeter per second, while the return is 1.293. This is because a lot of mass will be left at Mars. Outbound flight time is 605 days; return time is 660 days. The return leg is longer even though the acceleration is greater, because the geometry of the planets is not right. Note that the return trajectory gets as close to the sun as 0.6 astronomical unit—inside Venus's orbit.

A spacecraft designed for a Mars sample return mission

should be able to carry the entry vehicle, landing equipment, ascent rocket, a lander, and surface operations equipment. A JPL design study several years ago considered a Mars sample return mission and concluded that a direct return mission (using solar sail to get there and conventional ballistic for a direct return flight) would require a launch capability of from 7.5 to 10 tons, depending on launch year and· the types of propellant used. The more flexible orbital entry–orbital rendezvous mission design discussed above would require about 6.5 tons launch mass.

Thus, a sail with an area of ¾-million square meters—that is, either an 850-by-850 meter (a half-mile on a side) square sail or a 12-blade 6 kilometer-by-10 meter heliogyro of the technology assumed by JPL in the Halley mission design (0.1 mil plastic, sail vehicle loading of 4.8 grams per square meter) could be used for a high-performance Mars sample return mission. The landed mass at Mars is 3,000 kilograms, the interplanetary shuttle craft itself has a mass of 1,600 kilograms. An advanced technology sail—one that might be constructed in space—could propel a vehicle with a density of only 3 grams per square meter. This sail would permit landed masses of more than 4,000 kilograms for the same size sail. It is clear, then, that the present technology sail is adequate to initiate the interplanetary shuttle and the advanced technology sail makes it truly a remarkable bus for the solar system.

What about manned Mars missions? In 1981 Robert Staehle has performed the only analysis I know of such a manned planetary mission using a solar sail. He envisions a mission design with 14 solar sail cargo vehicles (with a sail 2 kilometers on a side and 0.1 mil thick) and 53 shuttle launches. The required launches take place over a four-year period. The mission would take about three years, from launch to return to Earth. The crew could spend one to three months on Mars. Mars aerocapture is assumed. Mars aerocapture uses the drag, caused when the spacecraft goes into the Martian atmosphere, to slow the spacecraft down so it can reach orbital speed around Mars. The sail returns to Earth orbit for subsequent use. Other new technologies were not assumed. It seems to me that now they should be. After all, this mission isn't going to fly for a few years. What are they?

Let's consider ultrathin—1-micron (0.04 mil)-thick—sails fabricated in space. The sail vehicle weighs about 5.5 grams per square meter. A 4 million square-meter sail (2 by 2 kilometers or 40 blades, each 10 kilometers by 10 meters) could carry a full shuttle payload to Mars, with a characteristic acceleration of 0.6 millimeter per second per second, in about 17 months. Utilizing aerocapture at Mars and manufacturing the propellants needed on-site at Mars with indigenous materials, this mission may need only 20 shuttle launches. It will need far less from a larger (heavy-lift) expendable launch vehicle, which one hopes the United States or the Soviet Union will have developed by the time this capability is needed. The mission can probably be done with little other advanced propulsion technology. Manned missions to Mars are not a near-term possibility, and the first such mission probably won't use a solar sail. As we noted earlier, the development of a low-thrust capability is economical only for regular use, not just for a single mission. Regular trips to and from Mars will have to be economical, and so the solar sail seems to be a real possibility for this kind of travel.

Another sail concept for human-crew missions comes from the inventors of the heliogyro sail. In one of their earliest reports they came up with a design for a sail vehicle to fit in a *Saturn V* booster, the huge rocket used in the *Apollo* program to launch men to the moon. Two payload capsules are located halfway between the center and the tip of the blades, getting the benefit of artificial gravity for the passengers from the centrifugal force (rotating acceleration!). The size of the deployed vehicle is prodigious—46 miles from tip to tip. It has 80 blades, each 23 miles by 10 feet. The total sail area is just under 100 million square feet—2,500 acres. The sail is 0.05 mil thick, and the total vehicle weighs 100,000 pounds (50 tons); half of that weight is sail. This design represents a different means of handling huge payloads in space. This sail could make a round trip to Mars in about three years. Human missions to Mars are discussed in the next chapter.

8

EXPLORING

OTHER

WORLDS

In November 1980, *Voyager 1* flew through the saturnian system and returned high-resolution images of the planet, its rings and large moons. Six of the planet's fifteen known moons appear in this montage compiled from *Voyager* images. Moving clockwise from the right, this view shows Tethys and pockmarked Mimas in front of the planet, Enceladus in front of the rings, Dione in the forefront on the left, Rhea off the left edge of the rings, and Titan in its distant orbit at the top. After swinging by Saturn, *Voyager 1* set out for the edge of the solar system. (JPL/NASA)

Now that we have seen how solar sails can take us through the solar system and to other planets, it's time to look at where we want to go and what we want to send there. For if we are going to build solar sails they will have to help in our exploration of the solar system. The first two decades of the space age were a remarkable time for those of us living on planet Earth. Each year between 1962 and 1981 we obtained pictures of new worlds, other worlds besides ours in the solar system, sent to us from robotic spacecraft visiting these planets. The moon, Mars, Venus, Mercury, and the forty-three remarkable worlds of the Jupiter, Saturn, and Uranus systems were displayed to us from the spacecraft in an increasingly spectacular array of photographs and other scientific data. Twenty-nine spacecraft went to the moon in the 1960s, another thirteen in the 1970s. Four traveled to Venus in the 1960s, nine in the 1970s. Three flew to Mars in the 1960s, another seven in the 1970s. The 1970s also featured one fly-by of Mercury and four of Jupiter. The total was thirty-seven planetary encounters in the 1960s, and thirty-two in the 1970s. Sixty-nine voyages of discovery in two decades!

What did we learn from our exploration? Mars and Venus are totally unlike each other, and both are very different from Earth. Yet the terrestrial triad—Venus, Mars, Earth—is beginning to reveal a fascinating story of our planet's evolution and the forces that control that evolution.

MARS

The extremely thin atmosphere and lack of liquid on its surface have prevented erosion from altering the martian surface very much. Therefore, our study of the Mars surface reveals to us billions of years in geologic history. By looking at these pictures, we can tell what has caused the climatic changes over history, and we can deduce from those causes what effects might have been presented on Earth during the same periods. Climate is, of course, controlled by the sun and, specifically, by the sun's output of energy, which "hits" the planets. Whatever caused climatic modifications on Earth also had effects at Mars. Earth is, of course, covered with water and has a heavy atmosphere, which has eroded much of the planet's surface. In addition, human modification has obscured much of the geological record. On Mars we have that geological record and we can "look back in time" to try to understand the history of the planets.

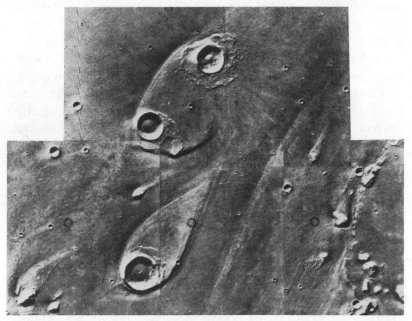

FIGURE 8.1 This mosaic of five Mars pictures shows the eastern part of the Chryse region near the prime Viking 1 landing site. The Viking orbiter cameras took the pictures from a range of about 1,600 kilometers (992 miles) in 1976. Braided channels record water flowing on the planet in the past. Fine grooves and hollows are on the upstream side of flow; obstacles also are seen. Shore of the channel is at lower right.

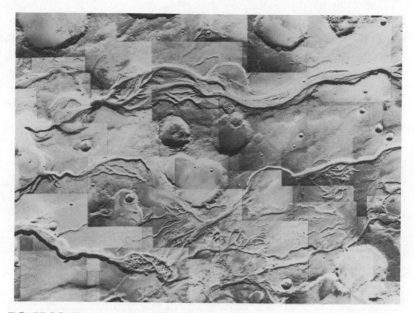

FIGURE 8.2 This is a mosaic of the lower portion of a complex of channels on Mars called Mangala Valles. The channels appear to have been cut when large amounts of water were released from beneath the surface. Visible are many streamlined islands and interior channels, typical of erosion by rivers and streams. Also visible are many examples of sediments originally laid by ponding and then recut by later flows. An example of chaotic terrain can be seen at upper right. Scientists believe it is a result of withdrawal of ground water or ground ice during a melting period. Some impact craters in the scene have been highly modified by erosion and wind-deposition of the surface material. The changes appear to be related to the process that formed the valleys. Other craters in the area are pristine and, therefore, relatively younger. Smaller valleys, some formed by the movement of materials within pre-existing, larger channels, can be seen cutting through the scarp or cliff at extreme right. There may be large amounts of water-ice still held in the surface materials in areas like this where flooding has occurred. (JPL/NASA)

Pictures sent to Earth by the *Viking* orbiter (Figures 8.1 and 8.2) show two of the most fascinating climate records so far discovered on Mars. The first shows certain evidence of running liquid on the surface. Is it water? Most scientists are convinced it is. If so, then either Mars was at one time a lot warmer or the martian atmosphere was much denser, because liquid water cannot exist on the surface under present conditions. If Mars did have a more benign environment in the past, it might have been a suitable place for life to start. In fact, it would have been much like primitive Earth. Was there life at one time on Mars and can we find fossil remains of it? Or has there never been life on that planet—and why? Either answer will be important for understanding the nature of life.

Mars today is very different from what it was aeons ago. At noon near the martian equator, it now gets just warm enough for ice to melt—but barely. The atmospheric pressure at the martian surface is about one-hundredth that of Earth—about like the pressure at a 100,000-foot altitude above the Earth's surface, which is higher than airplanes fly. This atmosphere is primarily carbon dioxide. Water cannot exist as a liquid in these conditions. Most of the water, if it exists in any significant quantity, is believed to be permafrost. For the last 3.5 billion years, Mars has been an arid and cold place.

A great deal of attention is being directed to the polar regions of Mars (Figure 8.3). The planet's south pole area is covered with frost composed mostly of carbon dioxide; the north polar region is believed to be composed mostly of water ice. The layered polar terrain provides a record of climatic history, because the conditions at different times have been frozen into different layers and we can examine all these layers. Ah, for the day when we can actually walk on this terrain, take samples of it, and make measurements! Since this is a very inhospitable area—where the temperature is usually −120°C (−184°F), where the wind blows at 200 kilometers per hour (120 miles per hour), and where the surface is like dry ice—we may have to do our exploring with robots—but maybe not.

Scientists are also studying martian global dust storms. The dust on Mars periodically forms a storm that darkens the planet, providing us with data that have helped us learn what the effects of nuclear war might be on Earth's atmosphere. A thermonuclear war could form huge amounts of dust, debris, and particulate matter which would become part of Earth's atmosphere, darkening our planet just as the dust storms darken Mars. Our new understanding of the catastrophic ef-

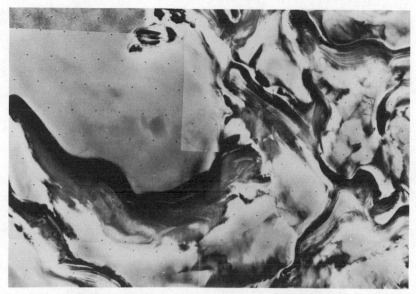

FIGURE 8.3 In this mosaic, compiled from images returned by the *Viking* orbiters, distinct layered deposits are visible within valleys cut into Mars's north polar cap by erosional forces. Each winter carbon dioxide freezes out of the atmosphere to form a seasonal polar cap; at the north pole, at least, the carbon dioxide frost overlies a residual cap of water-ice. The alternating light and dark layers are believed to reflect the differing rates at which frost and dust were deposited in the polar caps. This layered terrain at the martian poles may record climatic cycles and changes over the past few million years. (JPL/NASA)

fects of nuclear war may help us avoid such a disaster. This knowledge is a legacy of the *Mariner* and *Viking* missions to Mars.

VENUS

Our studies of Venus (Figure 8.4) have also increased our understanding of Earth. When scientists looked at the data from the first few atmospheric probes of Venus, they began to wonder what the human modifications to Earth's atmosphere would do to our own planet. On Venus they found aerosols, sulfuric acid content, excess carbon dioxide, and tremendous

FIGURE 8.4 In visible light, Venus appears as a featureless, yellowish ball. But when seen through an ultraviolet filter, as in this image taken by *Mariner 10* in 1974, details in the upper atmosphere begin to stand out. Thick sulfuric acid clouds enshroud Earth's nearest planetary neighbor; we knew little of our "sister" planet until the spacecraft explorations of the last three decades. Beginning with *Mariner 2* in 1962, a series of spacecraft from the United States and the Soviet Union have revealed a world very different from Earth. A crushing carbon dioxide atmosphere generates a runaway greenhouse effect that has heated the surface temperature to 460°C—hot enough to melt lead. In the dense atmosphere, the strong venusian winds would flow more like deep ocean currents than gentle summer breezes. Some scientists believe that erupting volcanoes generate lightning that crackles across the seething sky. Venus does not appear to be a pleasant place to visit. (JPL/NASA)

heat and aridity. Aerosols and sulfur are two major pollutants that humans have been introducing into Earth's atmosphere, and carbon dioxide increases are of great concern as we increase our use of fossil (carbon-based) fuels and strip huge land areas of their vegetation. Will Earth, too, evolve into a hot and arid desert? The "runaway greenhouse" on Venus was caused by the fact that hydrogen and oxygen in the primordial venusian atmosphere never had a chance to combine to make water. This is true because Venus is closer to the sun and therefore hotter than Earth. The water molecule was split apart by the sunlight. This split is called *photodissociation* and occurs when photons of sunlight hit water molecules hard enough to split them into hydrogen and oxygen. Without water, of course, there could be no rain; without rain, oceans could not form; and without oceans, the natural carbon dioxide of the atmosphere could not precipitate—that is, form sediments that we know as carbonates. (The carbonate sediments that formed on Earth were the result of the carbon dioxide precipitating in the ocean water.) With nothing to absorb it, the carbon dioxide blanket on Venus further insulated the planet, preventing the surface heat from escaping. This made Venus even hotter, which made formation of water even more difficult, and created the runaway greenhouse. It is called a greenhouse because the carbon dioxide acts like the glass walls and roof of a greenhouse, trapping the heat at the planet's surface.

By looking at Venus, we became more sensitive to Earth's atmospheric problems, and our studies of that planet continue to give us knowledge that will help us deal with present environmental forces and those that caused Earth's evolution. One of the most interesting studies now under way is the radar mapping of Venus. The venusian topography shows evidence of geologic forces at work. Understanding those forces will help us understand certain forces on Earth, including volcanoes, earthquakes, and storm formation. Figure 8.5 is a Soviet radar picture of Venus. The study of aerosol, sulfur, carbon dioxide, dust, nuclear explosion, and other such effects on Earth's atmosphere have been led by space scientists involved in planetary exploration missions.

JUPITER AND SATURN

Two mini-solar systems have been found at Jupiter (Figure 8.6) and Saturn. The worlds of these solar systems are very

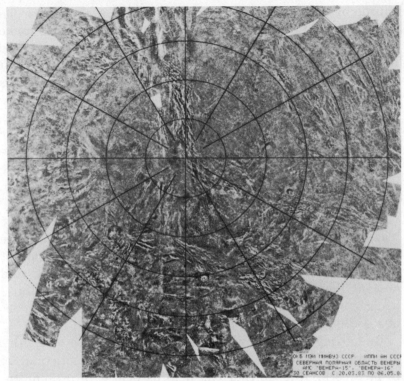

FIGURE 8.5 The Soviet Union has carried out an ambitious program of Venus exploration. They have landed craft on the surface, sent balloons floating through the atmosphere, and placed radar mappers in orbit. This mosaic of the north polar region was compiled from radar data returned by the *Veneras 15* and *16* orbiters. The surface relief seen by the *Veneras* appears to be the result of tectonic forces; some mountain ranges resemble the Appalachians of the eastern United States. (Vernadsky Institute of Geochemistry and Analytical Chemistry, USSR Academy of Sciences)

different from each other. At Jupiter, the giant Galilean satellites range from Callisto, which is cold, heavily cratered, has an ancient surface, and is dead, to Io, which is hot, active, and has a relatively young surface. The history of this system is graphically revealed to us when we look at the four satellites together. Figure 8.7 shows this. Io has a hot, active interior because of its proximity to Jupiter, which causes internal tides on the planet. The inside of the planet Io is not solid. It is mushy enough to be moved around, just as Earth's oceans are moved by the gravity of the moon and sun. In other words, Jupiter's gravity causes tides in the interior of Io. Because Io is hot and active inside, the sulfurous material in the planet keeps erupting through the crust. Europa, another planet in Jupiter's system, has much less internal activity. It is somewhat farther

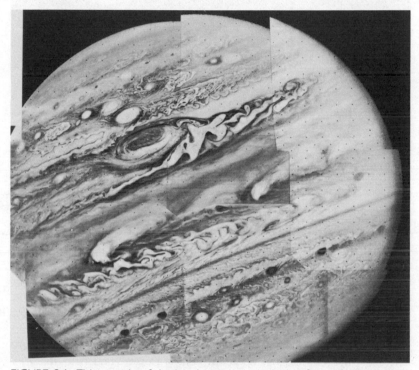

FIGURE 8.6 This mosaic of Jupiter has been assembled from nine individual photos taken through a violet filter by *Voyager I* on February 26, 1979. At the time, the spacecraft was 4.7 million miles (7.8 million kilometers) from Jupiter, heading toward a March 5 encounter. Distortion of the mosaic, especially noticeable where portions of the limb have been fitted together, is caused by rotation of the planet during the 96-second intervals between individual frames. The large atmospheric feature just below and to right of center is the Great Red Spot. The complex structure of the cloud formations seen over the entire planet gives some hint of the equally complex motions in the *Voyager* time-lapse photography. The smallest atmospheric features seen in this view are approximately 85 miles (140 kilometers) across. (JPL/NASA)

FIGURE 8.7 These photos of the four Galilean satellites of Jupiter were taken by *Voyager 1* during its approach to the planet in early March 1979. Ganymede (bottom left) and Calisto (bottom right) are both larger than the planet Mercury; Io (top left), and Europa (top right) are about the same size as Earth's moon. Image processing also preserves the relative contrasts of satellites. Thus, it is apparent that Europa has the least contrast; Io the greatest. Io is covered with active volcanos and a surface composed largely of sulphur. Europa is very different; *Voyager 1* did not approach Europa closely enough to show its surface in great detail. Ganymede and Callisto are both composed mostly of water and water-ice; they have large quantities of ice exposed on their surfaces. The Io photo was taken from 1.8 million miles (2.9 million kilometers); Europa, 1.8 million miles (2.9 million kilometers); Ganymede, 2 million miles (3.4 million kilometers); and Callisto, 4.3 million miles (6.9 million kilometers). Resolution of all photos except that for Callisto is about 30 miles (50 kilometers), and for Callisto it is 60 miles (100 kilometers). (JPL/NASA)

from Jupiter, but it has been active enough so that material from the inside has flowed across the surface. That material is now frozen into a fairly smooth crust. Ganymede, still farther out, has little of this activity, but it has many craters resulting from the billions of years of bombardment by solar system debris. They have not been washed over. The fourth planet, Callisto, is totally dead, and it seems as if every crater ever formed there is still revealed to us.

Planetary geologists describe surfaces as young or old. A young surface is one that was formed recently, perhaps as a result of eruptions washing over or covering the previous crust. An old surface is one that was formed aeons ago and that has not been changed lately. The early solar system had a great deal of debris in it, left over from its formation. This debris collided with the new planets and moons until most of it was swept up by these bodies. Naturally this accounts for a lot of collision-impact craters on the planets and moons. You can see them on Earth's moon and on Mercury and Callisto—old surfaces where the craters have been left undisturbed because these bodies are relatively dead. On other bodies, like Io, Europa, Venus, and Earth, the craters have been covered over by active geological processes such as volcanism and continent formation, so we know that these surfaces are younger. Older and younger regions sometimes exist on the same planet—the moon and Mars, for example—because of variations in local conditions.

The satellites of Saturn are varied and form a complex and interesting system about the ringed planet. There are 17 known satellites, eight of which were recently discovered by spacecraft. Saturn's moons interact gravitationally with one another, affecting not only their orbits and rotations but even the rings of Saturn. Interesting surface features resulting from impacts—probably from debris around the planet at the time the system was formed—are seen on six major satellites: Mimas, Enceladus, Tethys, Dione, Rhea, and Iapetus (see figure on page 81). The surfaces of these moons also show geological activity and evidence of hot interiors that cause volcanoes or other surface flows. But the one satellite that dominates the Saturn system is still the most mysterious: Titan.

Because Titan is cloud-shrouded, like Venus, we have never seen its surface, but we have measured its atmosphere. We know that the surface pressure on Titan is 60 percent greater than Earth's, and like Earth's, the atmosphere of Titan is composed primarily of nitrogen. Titan has haze layers (see Figure 8.8) at the top of its atmosphere and lower down, at about 300 kilometers, made by photochemical smog. Complex hydrocarbon and organic molecules are present at temperatures as "high" as −110°C (−166°F). According to one recent theory, the

FIGURE 8.8 Like that over the city of Los Angeles, a layer of photochemical haze enshrouds Titan, the largest moon of Saturn. Details in the layers stand out in the lower left section of this image taken by *Voyager 1* on November 12, 1980 from 22,000 kilometers (13,700 miles). Divisions in the haze appear at altitudes of 200, 374, and 500 kilometers (124, 233, and 310 miles). The thick aerosol clouds riding above the moon's limb can be seen in the upper right section of the photo. (JPL/NASA)

surface of Titan may be a liquid ocean. Its main liquid components are probably ethane and methane. Imagine a planet with a gasolinelike ocean (luckily there is no oxygen or it could burn up) and organic molecules in a smog layer precipitating out and raining onto this ocean. This is one ocean on which I don't want to sail but about which I must know more. Organic compounds are generally thought to be the precursors of life. Might the ocean on Titan be a cornucopia of organic chemistry from which we could learn about the origins of life?

The rings of Saturn are perhaps an even more exciting discovery. Before *Voyager*, most of us thought that Saturn had five or six rings (Figure 8.9), but the *Voyager* pictures revealed hundreds (Figure 8.10) and then thousands (Figure 8.11) of rings. The electromagnetic forces around Saturn and/or the gravitational interactions of the Saturn satellites with the rings make curious designs, such as spokes in the rings and twisting or braiding of some of the ringlets (Figure 8.12). Some popular press accounts claimed that these anomalies violated the laws of physics, but of course they don't. They simply show that we have an incomplete understanding of the effects of those laws. By seeing the new patterns in nature, we have been able to

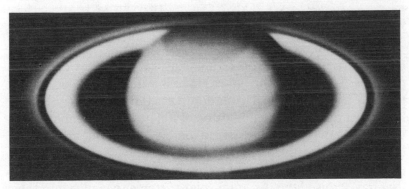

FIGURE 8.9 Even in the best pictures taken from Earth, few features are discernible within Saturn's rings. The 2,000-meter resolution of this photograph, taken from the Catalina Observatory of the University of Arizona, is enough to show (from the outside moving in) the A-ring, Cassini's Division, the bright B-ring, and the faint C-ring. (Catalina Observatory, University of Arizona)

FIGURE 8.10 As *Voyager I* approached Saturn, never-before-imagined detail began appearing in its images of the rings. Approximately 95 individual concentric features can be discerned in this two-image mosaic taken on November 6, 1980 from 8 million kilometers (4.5 million miles) away. An intricate interplay between Saturn, its satellites, and the ring particles generates the extraordinarily complex structure. Dynamical theorists will be kept busy for years trying to explain it. (JPL/NASA)

learn more about the forces and natural laws that cause the patterns. Sometimes this knowledge is expressed as one more term in a mathematical equation. Other times it may be the finding of a new celestial body or of a new phenomenon makes us realize that this new body or phenomenon affects the behavior of other planets or moons.

We don't yet understand all of the forces and interactions that have caused the two mini-solar systems to evolve into their present state. But by studying the new data, we are learning a great deal about the origin and evolution of planetary systems, the formation of planetary surfaces, and the dynamics of motion in the solar system.

FIGURE 8.11 As *Voyager 2* flew 743,000 kilometers (464,000 miles) from Saturn's rings, its camera revealed the incredible intricacy of the system. The pre-*Voyager* concept of four or five rings circling Saturn was drastically altered by images like this, which shows a small region of the thick, opaque B-ring. The narrowest feature seen here is about 15 kilometers (10 miles) wide. The brightness variations are due to a combination of differences in ring-particle density and light-scattering properties. (JPL/NASA)

URANUS

In January 1986, four days before the *Challenger* explosion, *Voyager* provided an extraordinary triumph of engineering ingenuity and performance. Eight and one-half years after launch, four and one-half years after its planned mission end (at Saturn), 1 billion miles from Earth, and with a crippled scan platform and a damaged communications link, the spacecraft sent back remarkable pictures and data about seven "new" worlds. These pictures showed us that Uranus and its satellites (Figure 18.13) were very different from what we had imagined—testimony to the ability of the universe to continually surprise and reward its explorers.

FIGURE 8.12 *The kinky F-ring was one of Voyager I's strangest discoveries. Two small "shepherding" satellites orbit Saturn on either side of this ring, and scientists believe that gravitational interactions among the satellites and the ring particles create these strange structures. (JPL/NASA)*

While looking at photographs of Miranda, one of Uranus's moons (Figure 8.14), planetary scientists commented on how bizarre the satellite looked—as if it contained all the geology ever dreamed about on a single body in the solar system. Uranus's magnetic field was even more bizarre—clearly the result of a tumultuous past that caused the strange rotation of Uranus and its magnetic field. The discoveries at Uranus offer us new opportunities to analyze the physics of the solar system.

WHAT NEXT?

So it was a remarkable two decades, with new worlds revealed and with new knowledge gained, knowledge that we have used to understand our own planet as well as the forces that surround us in the universe. However, the process of exploration so wisely begun in the early 1960s almost came to a complete

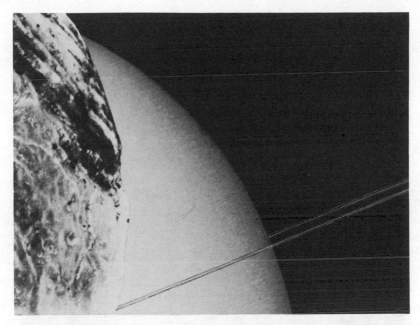

FIGURE 8.13 A future visitor preparing to land on the icy surface of Uranus's moon Miranda might witness this stunning view of the planet's cloudtops, some 105,000 kilometers (65,000 miles) away. This montage of *Voyager 2* images obtained in January 1986 shows Uranus overlaid with an artist's conception of the planet's dark rings as they might appear to the lucky tourist. A portion of Miranda, seen in a *Voyager* close-up, fills the foreground and shows the view along one of the huge canyons scarring the moon's surface. (JPL/NASA)

stop for Americans in the early 1980s. At the end of the 1970s, the United States experienced a systematic year-by-year scale-down of our solar system exploration plan. Only one U.S. mission was planned for the 1980s, and the new Reagan administration even made an attempt to call off the already flying *Voyagers* and have them stop taking data after they passed Saturn. The ridiculousness of this was quickly seen and the idea forgotten. Then, in 1982, for the first time in 20 years, no pictures came back from a new world. This situation has continued for several years.

Was the United States, like so many earlier civilizations, going to begin a process of exploration and then quit? The

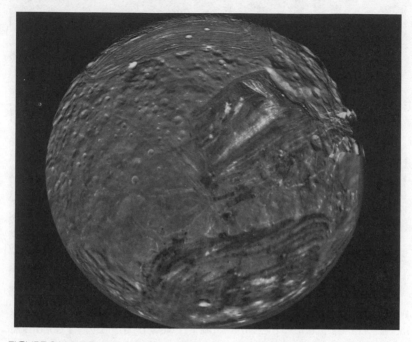

FIGURE 8.14 "Miranda is a bizarre hybrid . . . of valleys and layered deposits on Mars, combined with the grooved terrains of Ganymede, matched with (what some have called) compression faults on Mercury. So if you can imagine taking all the bizarre geologic forms in the solar system and putting them into one object, you've got it in front of you." Thus did Laurence Soderblom of the *Voyager* Imaging Team introduce Miranda, the innermost and smallest of the main moons of Uranus. Although it is only 500 kilometers (310 miles) in diameter, Miranda displays some of the strangest geology yet seen in the solar system. Embedded in cracked crust and surrounded by gently cratered, rolling hills, a chevron-shaped feature dominates the center of this mosaic. Two concentric ring features on either side of the chevron resemble racetracks and were whimsically dubbed the "Circi Maximi" by the Imaging Team. (JPL/NASA)

Vikings, the Celts, the Chinese, the Portuguese, and others all began exploration of the globe and then ceased, leaving the job to others. Their reasons for stopping always had something to do with a political crisis or internal problems. History judges those decisions harshly, for it is difficult for us to care about those internal problems, but we do remember well the exploration endeavors.

Other countries are now taking up the challenge of exploring the solar system. After the United States backed out of its commitment, the European Space Agency (ESA—a consortium of European nations) decided to go ahead with the International Solar Polar Mission, now called *Ulysses*, to fly a spacecraft over the sun's pole to examine how the sun behaves as a star. ESA also decided to launch a fly-through of Halley's Comet, as did Japan. And the successful Soviet program, which has sent several landers to Venus, became more ambitious and, incidentally, more international. The Soviets also decided to restructure a planned Venus mission into the Venus–Halley mission known as *VEGA*.

HALLEY'S COMET

The *VEGA* mission was possible because the position of Venus in its orbit was such that a spacecraft from Earth could go around that planet and then proceed to Halley's Comet in a relatively short time without using propulsion to change the orbit. This was the first application of gravity assist by the USSR. The Russians have conducted a nine-nation mission with some 16 experiments, including a balloon into Venus's atmosphere, a lander on its surface, and then the fly-through of Halley's Comet in 1985–86. *VEGA* was an extraordinary mission involving several Eastern European nations, West Germany, and France. There were even three American co-investigators on the *VEGA* mission team. The openness of the planning and development of the mission, and the broad international involvement, were departures from previous missions. It represented a new confidence by the Soviets in space exploration and offered new opportunities for science and exploration for everyone.

On its fly-through of Halley's Comet, the spacecraft aimed for a passing distance of approximately 10,000 kilometers or less. A dust shield was designed to protect the spacecraft against debris that flies off the comet as it nears the sun, the result of the sublimation of the comet's nucleus, which is probably composed of dirty ice and snow. *VEGA* obtained the first close-up pictures (Figure 8.15) of a cometary nucleus, and *VEGA* sampled and analyzed the comet's particles. It also caught the molecules as they came off the nucleus and determined their composition before they were chemically affected by the interaction with the solar wind—that stream of particles and ions that the sun blows out through the solar system. These "parent" molecules may turn out to be our best chance of sampling material from the time of the solar system's formation. The *VEGA* mission also measured other properties of the comet, including its magnetic field and the distribution of charged particles near the nucleus and through the coma, or atmosphere, of the comet.

A couple of days after the Soviet mission in March 1986, the European Space Agency's spacecraft, *Giotto*, also arrived at Halley's Comet. *Giotto* was targeted right at the nucleus. Of course, the probability of hitting it directly was very small, but the ESA planners minimized their fly-by distance and took

FIGURE 8.15 On March 6, 1986, *VEGA 2* passed within 8,000 kilometers of the nucleus of Halley's Comet. The craft got a broadside view of the comet, and scientists were able to determine that the comet had an oblong shape. In this image, a jet of dust extends down from the nucleus. Both the *VEGA* and *Giotto* flew through such jets, and many instruments were damaged by dust impacts. (Institute for Space Research, USSR Academy of Sciences)

close-up pictures of the comet's nucleus (Figure 8.16 and 8.17). In an extraordinary example of international cooperation, the USSR, the ESA, and the United States agreed on a navigation plan for the Halley spacecraft. Soviet pictures were made available as navigation data for targeting the *Giotto* mission. Also, the United States cooperatively tracked the Soviet spacecraft to provide additional navigation information and reduce the error of the spacecraft's position. The ESA spacecraft also made other measurements, obtaining samples of the particles near the nucleus and the particles in the comet's coma.

The Japanese spacecraft, *Suisei* ("Comet"), passed approximately 100,000 kilometers away from the comet. It provided remote images of the comet's nucleus with an ultraviolet sensor.

It is ironic that the engineering feasibility of the solar sail

FIGURE 8.16 With images like this one taken by *Giotto* on March 13, 1986, scientists were at last able to take a close look at the heart of a comet. Here sunlight is coming from the left, heating the icy nucleus so that dust jets erupt from its surface. The jagged edge of the terminator (the line between day and night) indicates that the nucleus is very irregularly shaped. Near the top are bright spots, and below them is a craterlike feature about 1.5 kilometers in diameter. This could be a slump feature formed by outgassing of nearby jets. Below and right of the "crater" is a bright feature that some scientists call "the mountain." (Image taken by the Halley Multicolor Camera, © 1986 Max Planck Institut fur Aeronomie, Lindau/Harz, F.R.G., with additional processing by H. J. Reitsema and W. A. Delamere of Ball Aerospace.)

was determined by a design for a planned U.S. rendezvous mission with that famous celestial visitor. But that rendezvous was denied, and no U.S. mission was initiated. NASA would not support a Halley's Comet mission of any type because it had other priorities. Both the Carter and Reagan administrations ignored public interest in the comet. All through the 1970s as we were conceiving new missions to extend our planetary exploration, we had been certain that in 1985 the United States would launch a mission to Halley's Comet. Other, more grandiose ideas were problematic (including the solar sail) but a simple probe to Halley's Comet seemed certain. After NASA called off rendezvous, it tried to develop a plan for a rendezvous with a much smaller, easier-to-reach comet named Tempel II. This plan called for a European Space Agency probe that would separate from the main spacecraft on the way to Tempel II and go to Halley's Comet. Thus, at least, the United States would be a partner in a Halley's Comet mission. But the Tempel II mission depended on the development of solar-electric propulsion, described earlier, and the solar-electric propulsion development plan was turned down. ESA decided to do the Halley's probe alone, with an *Ariane* launch vehicle. The United States decided to do nothing. Nonetheless, in 1986 the comet,

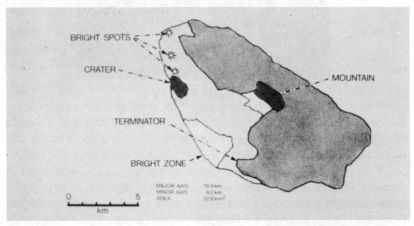

FIGURE 8.17 Details of the nucleus of Halley's Comet, see in Figure 8.15, are called out in this illustration. (S. A. Smith, from a sketch by H. J. Reitsema)

for the first time in its many-millennia history, was greeted by the people of Earth with an array of scientific instrumentation from four European and Asian spacecraft. Although the United States alone among the spacefaring nations chose not to send a spacecraft to Halley's Comet, it did participate with other nations in an International Halley Watch to coordinate the many space and Earth observations of the comet.

Space observations were made not only from the space mission flying through the comet, but also from Earth-orbiting spacecraft with telescope capability for looking at the comet, and from interplanetary spacecraft that can make complementary measurements—for example, of the solar wind and interplanetary fields and particles during the time the comet is in the inner solar system. The United States also redirected an already-flying spacecraft—the International Sun–Earth Explorer, which was renamed International Comet Explorer, ICE—to fly near a much smaller comet, Giacobini-Zinner, to learn more about comets. The United States also planned to conduct a shuttle-space lab mission to observe Halley's Comet. This was lost in the aftermath of the *Challenger* accident. The International Halley Watch is organized into teams of scientists built around the special disciplines connected with cometary science. Their job is to analyze, catalog, and explain the data that were taken during this passage in 1985–86.

By 1990, data from the Halley Comet fly-through missions will be analyzed. A rendezvous, which would permit a long-duration flight in formation with another comet, would be a logical follow-on. A rendezvous would enable scientists to make detailed observations of a short-period comet, thus gaining important new information about this class of objects. The rendezvous being considered is with Tempel II, which has a period of five years and will have a close approach to the sun in September 1999. The *Mariner Mark II* spacecraft is a free-flying ballistic machine, being planned for a new generation of U.S. missions—the first of which is a rendezvous with the comet. The mission is tentatively named *CRAF*, Comet Rendezvous–Asteroid Fly-by, since a fly-by of a main-belt asteroid is planned on the way out to the rendezvous. The spacecraft

(Figure 8.18) will approach the comet when it is well out in its orbit, before it begins to get active. The spacecraft will then fly in formation with the comet through the inner solar system.

GALILEO TO JUPITER—SOMEDAY

The *Galileo* mission to Jupiter was supposed to be launched in 1986. (As of June 1987, the launch is scheduled for November 1989.) The lack of U.S. launch capability after the *Challenger* explosion prevented that. *Galileo* will carry the most ambitious payload ever sent into interplanetary space. The spacecraft has two parts: a 2,000-kilogram orbiter carrying nine instruments, and a 335-kilogram atmospheric probe with six instruments. Before reaching Jupiter, the spacecraft will release the probe to penetrate the gaseous giant (Figure 8.19) and radio back to the orbiter information about the composition, temperature,

FIGURE 8.18 As it is now planned, the Comet Rendezvous–Asteroid Fly-by (*CRAF*) spacecraft will be launched in early 1993. CRAF will swing out through the asteroid belt and fly by a large asteroid called Hestia 46. The craft will then move on to rendezvous with Comet Tempel II in October 1996. *CRAF* will join the comet when it is at its farthest point from the sun, and so at its least active, and then follow it in as it approaches the sun and develops the bright, tailed appearance that we associate with comets. This rendezvous mission will teach us about the life cycle of a comet—only briefly glimpsed by the *VEGA* and *Giotto* fly-bys. (JPL/NASA)

FIGURE 8.19 Sometime in the 1990s, a probe from the *Galileo* spacecraft will penetrate the atmosphere of Jupiter and give humanity its first taste of a giant gas planet. Slowed by a parachute and protected by a heat shield (being jettisoned in this illustration), the probe should sink deep into the jovian atmosphere, and may sample ammonia, ammonium hydrosulfide, and water clouds before it is crushed by the tremendous pressures of the interior. (NASA illustration)

and pressure of the atmosphere before the probe is crushed in the depths of the planet. After separating from the probe, the orbiter will brake into orbit about Jupiter and repeatedly swing past the large Galilean satellites, returning images and information on the planet and its moons.

Repeated fly-bys (Figure 8.20) of the Galilean satellites of Jupiter (Io, Europa, Ganymede, and Callisto) will be made possible on this mission by targeting one encounter so that the satellite perturbs the spacecraft's trajectory to the next encounter. John Beckman of JPL developed this idea into an orbital "tour" of the jovian system. He coined the term "orbit pumping" to describe the changes in orbital energy resulting from satellite encounters. Then Chauncey Uphoff and Phillip Roberts, also of JPL, came up with "orbital cranking," which would use gravity assist to change the orbital plane, allowing more tours

FIGURE 8.20 When it reaches the jovian system, the *Galileo* orbiter will use gravity assist to swing through the system again and again. Accurately targeted by mission controllers, the spacecraft will swing by Io and then repeatedly encounter Europa, Ganymede, and Callisto. (NASA illustration)

through the system. Mission designers will also use gravity assist to ease *Galileo* into Jupiter's orbit by flying it close to Io and losing energy before braking.

A fly-by of Io will be done only on this first orbit because the radiation that close to Jupiter is very dangerous. But the craft will make repeated close fly-bys of the other three major satellites. By targeting close to Europa, Ganymede, or Callisto, the trajectory can be modified so that the orbit is reshaped for the next encounter. In this way, the spacecraft's encounters with the three major moons can be used not only to make measurements at these moons but also to target the spacecraft for the next encounter with a different moon. After one year in orbit, *Galileo* should provide one dozen satellite fly-bys as well as the atmospheric data from the probe into Jupiter. In addition to making imaging measurements, the orbiter will carry out an extensive fields-and-particles survey as it traverses through different parts of the jovian magnetosphere. Figure

8.21 shows the evolution of the *Galileo* orbit during the mission.
The spacecraft will need a great deal of energy to get from
Earth orbit into an orbit around Jupiter. The mission was
planned as the first planetary launch on the shuttle. Even using
the powerful *Centaur* rocket as a stage out of the shuttle it
would have required several tricks of celestial mechanics to
complete *Galileo*'s jovian tour. These tricks are based on the

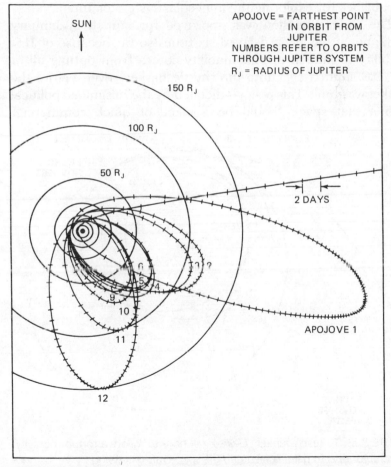

FIGURE 8.21 If all goes well, the *Galileo* orbiter could make 12 passes past the
large moons of Jupiter. (JPL/NASA)

principle of gravity assist (Figure 8.22). But the shuttle program has gone all wrong, and the Jupiter mission has yet to go. The shuttle capability for the 1989 launch now planned is much lower than the original 1982 launch plan, and now there is no *Centaur.*

The history of the *Galileo* mission is a microcosm of the history of what's gone wrong with the U.S. space program. Conceived in the mid-1970s, at the height of U.S. solar-system exploration, it is an ambitious extension of the *Voyager* capabilities for exploring the mini-solar system of Jupiter.

The *Galileo* mission was approved for launch in January 1982, but it has been delayed six times so far because of U.S. inability to launch it. This inability comes from putting all of the U.S. launch eggs into the shuttle basket—even before the basket was built. This policy—dictated by the misguided political notion that space should be a place of quick commercial

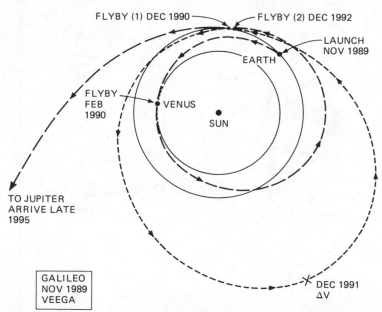

FIGURE 8.22 To reach Jupiter, *Galileo* will have to follow a tortuous gravity-assist trajectory. To gain momentum, the spacecraft will swing by Venus once and Earth twice, making nearly three orbits through the solar system. (S. A. Smith, *The Planetary Report,* adapted from JPL/NASA.)

benefit rather than a place of discovery and adventure—still dominates the U.S. program even after the *Challenger* disaster. Thus, the United States, once a major player in the exploration of the solar system, is now virtually benched. Ironically, we are also having a slowdown of the U.S. development of space capabilities even for applications to social needs—communications, land-use and climate observations, global environmental monitoring, navigation, etc.—because money is being redirected to the big shuttle and space-station projects in pursuit of engineering goals, not scientific ones.

RADAR MAPPING OF VENUS

The United States was to have launched a radar mapper to Venus in 1988 with an arrival at Venus in 1989. The mission is now scheduled for the next Venus opportunity—19 months later. The Venus radar mapper, now called *Magellan*, should be able to follow up the 1983 Soviet mission aimed at mapping the rest of the planet and overlapping areas at higher resolution. The Soviet Union's *Venera 15* and *Venera 16* entered orbit around Venus in October 1983, to the surprise of U.S. scientists. Radar mapping of Venus had been a goal of the U.S. program since the early 1970s when familiarity with the technique of synthetic aperture radar was becoming known to the civilian scientific community. Prior to that it had been solely a military tool for all-weather reconnaissance from airplanes. The idea is to use the direct measures of distance that the radar signal provides. This is done by measuring the time it takes a signal to be broadcast and then bounce back from a target. At the same time one uses the indirect measure of velocity obtained by the Doppler effect—that is, a change in the bounced-back signal's frequency, which comes from the relative motion of the transmitter and the target. The distance and the velocity give two dimensions of information that can be translated by data-processing techniques into an image. This image is different from a photographic image, but it gives the same kind of information. The pictures in Figure 8.5 are radar images of the Venus surface from the Soviet *Venera* spacecraft. The Soviet

Union mapped the northern part of Venus with about 1- to 2-kilometer resolution (the size of an object you can distinguish). The U.S. mission will map the southern, as well as the northern, half of the planet to about half that size. The combination of the American and Soviet radar mapping of Venus may give us solid evidence about the presence of active volcanism on Venus's surface. The mappings may also reveal whether the surface of the planet is composed of geologic plates like Earth's, which, as they moved (and continue to move) on the planet, triggered earthquakes and formed the continents.

BACK TO MARS

While the United States is exploring Venus with *Magellan*, the USSR will be conducting an extraordinary mission to Mars, where its spacecraft will rendezvous with the small martian moon Phobos. The Soviets have long been fascinated with Phobos and have even been known to suggest that it might be an *artificial* satellite, a hollow chunk of rock—perhaps a remnant of an ancient extraterrestrial civilization. Thanks to analyses of Phobos's orbital path and close-up views taken by *Mariner 9* and the *Viking* orbiters, however, we now know that Phobos is a natural object. The Soviet mission, planned for a 1988 launch, is really three missions in one: an orbiter to provide remote observations of Mars; an interplanetary solar wind fields-and-particles monitoring platform; and most significantly, a rendezvous with and a landing on Phobos. The detailed close-up observation of Phobos, with imaging of the surface down to 10-centimeter resolution, and determination of its chemical and mineralogical content will provide exciting information about the origin of this body, its relation to Mars, its composition and evolution, and its possible significance as a future Mars outpost. Phobos looks like an asteroid. Did Mars capture it in some strange perturbation of celestial mechanics early in the history of the solar system? Is it like other asteroids or different from them? Was it formed with Mars or as a totally different body? The possibility of using it as a future outpost has been raised by several scientists including American geologist Fred Singer and former astronaut Brian O'Leary, with suggestions that landing human crews on Phobos—or on Deimos, the other martian moon—might be a sensible first step toward human exploration of Mars.

The Soviets have come up with an interesting and exciting method of measuring Phobos's composition. When the spacecraft gets within 50 meters of the moon, a laser fired at the surface will produce a vaporization at a small spot and release a cloud of plasma (atoms and ions liberated by the heat). The plasma will be observed and sampled so that the electrical charge and mass of the liberated molecules can be measured. This will permit scientists to deduce what the molecules are and, hence, what Phobos is made of.

The United States planned to launch a geoscience and climatology mission to Mars—the *Mars Observer*—in 1990. This was supposed to be the first of a new series of spacecraft adapted from commercially available Earth orbiters known as Planetary Observers. However, NASA's commitment to such a series is now very much in doubt. The Mars mission was delayed to 1992. It will have an outstanding imaging system that will be able to see objects 5 feet in size on the surface—the old *Viking* lander, for example. A complement of remote sensing instruments will attempt to inventory the planet's water, telling us where it is and how much there is of it. Gaining knowledge of the history of climate will also be an objective of the mission; the martian surface should reveal a record of its history. These measurements will be important to us when we plan the big steps of robotic and human exploration of Mars.

Both the United States and the Soviet Union have renewed their interest in the planet Mars. Scientific understanding of Mars is crucial to our understanding of the terrestrial planets and of Earth's environment. It is sobering to realize that we have visited Mars at only two places, and they are two of the most uninteresting sites that we could have chosen. Naturally, when we chose the landing sites for the *Viking* missions, safety had to be our prime consideration. But safety almost automatically implies dullness. The investigation of the enormously interesting polar regions, the layered terrain, the canyons, and those structures that appear to be ancient flood channels still awaits us.

We next need a surface exploration mission using a combination of robotic vehicles—remote sensing, *in situ* measurements and a sample return that can bring back pieces of Mars for analysis to determine the composition and age of different regions on that planet. As the *Viking* pictures have shown us, Mars cries out for further exploration. To understand the evolution of climate and water on that planet, to look beyond the horizon at regions where rivers apparently once flowed, and to look below the surface will all be a part of a very interesting and new mission for the exploration of Mars. The robotic and then human exploration of Mars makes sense both scientifically and economically. The development of a manned mission to Mars will take at least 15 to 20 years. For its engineering and for the science that we need to define the mission requirements, the knowledge from robotic precursor missions will be important.

The connections between understanding Mars and understanding the Earth are becoming clearer. As I mentioned earlier, the study of martian dust storms helped us understand the potential results of a nuclear war. Understanding the past course of liquid flow on Mars will tell us about the causes of climatic variations on that planet and, hence, on Earth.

We need a global survey of Mars. We need to know the surface chemistry and composition. We need samplings from many spots around the planet, and we need to explore under the rocks and into the channels of the surface. We need detailed analyses of surface chemistry and content of surface material. This means that we must develop a program of exploration that will include an extensive surface exploration of the planet, more surveys from altitude, and the return of samples for analysis. In addition, the construction of a scientific base for *in situ* analysis will clearly be necessary in the future, perhaps first with robots and then with humans.

A recent invention by Dr. Jacques Blamont, in France, could revolutionize the exploration of Mars. Blamont's invention is a dual thermal lighter-than-air balloon. The balloon is made of an ultrathin plastic sheet manufactured much like the solar sail. It uses hot air from solar heat during the day to rise and travel in the martian wind from one spot to another. At night it lands and is able to take measurements on the surface. The hydrogen- or helium-filled balloon always stays inflated; hence at night it holds the landed package taut so that it doesn't drag or snag on the surface. A valve on the thermal balloon also allows it to land in the daytime.

This balloon will give us a way to explore Mars and investigate interesting sites economically. The Soviet Union is planning to use it on a 1992 mission, which may also include a small rover that can travel along the martian surface. The rover is a wheeled vehicle that can rove around the planet's surface collecting samples, making measurements, and taking pictures. Another vehicle, known as a sample return, can go to the planet, pick up samples, and bring them back to Earth for analysis. We will need both types of vehicles if we are to explore the surface of Mars.

Many different types of rovers are possible. A list of them is given in the table below. The most common rover is similar to the one the United States and the Soviet Union used on the moon—a sort of stripped-down car that can work autonomously

TABLE 8.1   "Rovers" for Exploring the Surface of Mars

Vehicle	Advantages
Automobiles 　with wheels like cars 　with treads like tanks 　with legs for climbing	Moderate traverses, detailed surface measurements. Probably near autonomous.
Mini-Rovers, 　small wheels	Short traverses, lightweight and simple.
The Ball	Simple.
Balloon	Simple, highly mobile, long traverses, and wide coverage.
Airplane	Long traverses and wide coverage.

on the planet (see Figure 8.23). These vehicles could be set down on the surface, perhaps in pairs (so they could take pictures of each other), and they could rove around collecting samples and doing limited *in situ* analyses. They would have to carry a power system, probably a small radioisotope thermo-electric generator and batteries such as the one on the *Viking* lander, for locomotion and the scientific instruments. The out-standing feature of the rovers, increasing their complexity and cost, is near autonomy. These vehicles will also have artificial intelligence to help them cope with obstacles and steering problems.

An innovative idea has come from the French as an out-growth of their balloon program, namely, an inflatable ball that could function as a simple (and cheap) rover, rolling around on the surface and stopping to make measurements. The initial idea was extremely simple: Just land the ball on the surface,

FIGURE 8.23 The first rover to traverse Mars may be a sort of stripped-down car, with treads in place of wheels to carry it across the uncertain terrain. The craft will probably carry instruments to make some *in-situ* examination of martian materials. If it is part of a sample return mission, the rover could pick up rocks and soil, capture some atmosphere, and deliver the samples to a companion vehicle that would return them to Earth for extensive analysis. (JPL/NASA)

inflate it, and let it be blown around by the wind. However, such a device would quickly settle down into a depression or crater and never get out, thus ending its usefulness. And so a slightly more complicated ball was devised. It contains an internal system of weights that can be used to overturn the moment of inertia and push the ball in different directions. Students at the University of Arizona have investigated the concept, built a prototype, and come up with a feasible design for a Mars ball (Figure 8.24). In this version the need for complexity in design is overcome by the huge wheels that simply roll over things.

A very different type of exploration vehicle is the Mars airplane. The idea for this solar-powered airplane was conceived a few years ago by Dale Reed of the NASA Dryden Flight Research Center. He was responsible for building remotely piloted planes that worked at a 100,000-foot altitude

FIGURE 8.24 One of the more imaginative designs for a Mars rover is called the Mars ball. In its original incarnation, the Mars ball resembled a large beach ball; it would have rolled across the surface, stopping occasionally to pick up samples. Students at the University of Arizona came up with this version, which would use large, balloonlike tires to cross the martian landscape.

above Earth where the pressure is the same as the surface pressure of Mars. A conceptual design of a Mars airplane is shown in Figure 8.25. What a truly great exploration vehicle this would be, for it could fly many hundreds of kilometers a day, imaging great expanses of the martian surface, perhaps even landing and taking off again to provide some capability for surface sampling and experimentation. *Viking*-size engines would permit landing and takeoff capability, for if the solar-powered airplane is to last for more than one Mars day, these capabilities would be necessary.

A simpler Mars airplane design would provide one-day coverage, which could still cover an enormous amount of area on the planet. The airplane could be packaged neatly, like a

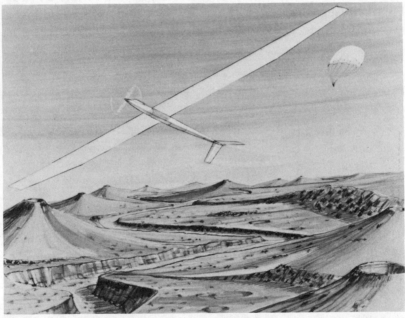

FIGURE 8.25 Perhaps someday a graceful, solar-powered airplane will cruise through the martian sky, gathering data for researchers back on Earth. Remote-controlled planes have investigated Earth's atmosphere at 100,000 feet, where the pressure is the same as the surface pressure on Mars, so it is technically possible to fly across Mars. (JPL/NASA)

cruise missile, and several of them could be carried to Mars on a single mission.

The airplane's extended survey capability and the ball's simple wind propulsion are merged in the concept of the solar balloon.

The design of a Mars surface sample return has been considered in various NASA studies. The detailed analyses of samples from Mars will provide information about evolution of the planet. These analyses will show us the mineral and chemical environment with which human explorers will have to cope. There are two basic concepts. One is to send the entire return vehicle to the surface of Mars, have it collect the samples, and then bring them directly back to Earth in the return vehicle. This involves sending a large payload to the planet, for it would have to include the return propellant. The other design concept calls for sending the sample return to Mars with just enough fuel to get it back into Mars orbit. Once in orbit, it would rendezvous with an Earth-return vehicle, which would bring the samples back to our planet.

One recent idea being developed by Professor Robert Ash at Old Dominion University is to use indigenous Mars materials to manufacture on the surface of Mars a propellant for return. Mars has oxygen, nitrogen, and carbon, so we could manufacture fuel robotically if we had time. Several approaches to fuel manufacturing on Mars have been suggested. The most straightforward seems to be to take oxygen from the atmospheric carbon dioxide; this is done by compressing it and then disassociating it into carbon monoxide and oxygen by electrolysis. The oxygen could be combined with methane, which would have to be carried up from Earth. It would be nice if we could avoid sending up anything from Earth—or if we could send only hydrogen, which is very light. But getting enough material on Mars may not be possible in the early days. At any rate, such schemes are not likely to be employed until we have much more experience on Mars.

Mars rovers and sample returns are vital developments for continued space exploration. We simply must know more about the red planet, for it represents the best, if not the only, hope for human settlement off our own planet. While the practical need for this is not yet known, the philosophical significance is obvious.

Where does solar sailing fit into these plans? As we have seen, the solar sail would provide an ideal vehicle for round-trip interplanetary transfers. Used as an interplanetary shuttle, the sail would enable us to collect samples on Mars and other planets on a continuing basis. With its continuous low thrust and its lack of need for propellant, it is ideally suited to round-trip missions. It could also enhance the other missions we have discussed. It could be used, for instance, for asteroid rendezvous, comet rendezvous, sample return, and delivery of larger payloads to Mars and the other planets. But for just one of these missions it wouldn't make much sense to develop a solar sailing vehicle. We don't need the technology to enable these missions, and the gain that would come from just improving them is not sufficient to justify the cost of building a whole new class of spaceships. This logic has prevented development of low-thrust (electric or sail) propulsion in the past. I think it is fair to conclude that solar sail vehicles for solar system exploration is an idea for the future. Small solar-sail vehicles for special purpose missions might, however, be used sooner, perhaps, as a lower cost option for a near-Earth asteroid rendezvous as proposed by the World Space Foundation (Chapter 10.)

Solar sailing will become really useful once regular and repeated expeditions to the planets are under way—when we've gone beyond the reconnaissance stage to the more detailed and intensive exploration of the solar system. The solar sail would allow us to deliver larger payloads, it would have the advantage of using no fuel, and it could be used repeatedly. Thus, it could become a very economical means of traveling around the solar system, perhaps as far out as Jupiter. Even beyond that distance where the sunlight is weak, the sail could be the basis of a continuing solar system transport system. This is seen in the next chapter.

9
INTERSTELLAR
FLIGHT

Three-stage laser-pushed lightsail. The inner two sectors pull away from the ring-shaped outer sail. Laser light from the solar system bounces off the ring sail, back onto the double inner stage, bringing it to a halt at the target star. (Painting by Seichi Kiyohara)

We are pretty pessimistic about the chances of finding life elsewhere in our solar system. It may be that there is some pocket of primitive microscopic life somewhere on Mars or perhaps on an asteroid or in a warm spot in the atmosphere of Titan or in the ocean of Europa—but it's highly, highly unlikely.

Conversely, we are beginning to accumulate evidence of the common existence of planetary systems and hence the likely existence of suitable abodes for life around other stars. We also see evidence that evolution of life on Earth was not all that lucky—that life should start and evolve when conditions are favorable. We need to see and study other star systems and see what happens at other places throughout the galaxy. Since our motivation is really a basic human characteristic—curiosity about our place in the universe—going to the stars is a natural goal to consider.

In this chapter we must extend both our concept and our definition of solar sailing. Without sunlight we can't have *solar* sailing, and between the stars there is, of course, no (or minuscule) sunlight. Space is mostly black, and mostly empty. To sail between the stars we will have to use artificial light—lasers.

But before we consider the light source and the mechanics of laser sails, we must first find out how difficult interstellar flight really is. The distances between stars are vast. The second-nearest star (Proxima Centauri) is approximately 4 light-years away from us. (The nearest star is the sun—a trick question in many astronomy courses.) A light-year is the *distance* (not the time) that light travels in one year—about 9.5 million million kilometers (6 million million miles). The speed of light is 300,000 kilometers per second, or 186,000 miles per second. Light travels from the sun to Earth in about eight minutes; from the sun to Pluto it takes light more than five hours. The distance from Earth to Pluto is about 0.0006 of 1 light-year or one-hundredth of one percent of the distance to the nearest star.

Let's dwell on this for a while, for if we are going to think about interstellar flight we'd better get comfortable with really big numbers. Six million million miles is written as 6,000,000,000,000 or, in scientific notation, 6×10^{12} (6 followed by 12 zeros). It is also called 6 trillion miles. To make an understatement, this distance is going to be hard to cover. The fastest vehicle now in existence is a *Voyager* spacecraft that is escaping the solar system after getting a gravity assist from Jupiter and Saturn. Its speed is about 30,000 miles per hour or 15 kilometers per second. At this speed, it will be at a distance of 4 light-years from Earth in 94,000 years! To travel that far in 100 years, our average speed (including acceleration and deceleration times) would have to be about 29 million miles per hour. This is about 4 percent of the speed of light.

An average speed of 29 million miles per hour over a hundred years requires us to start at, for example, zero, acceleration to 58 million miles per hour in the first 50 years, and decelerate back to zero in the next 50 years. (This just gets us there, not back.) The acceleration to be applied is over 1 million miles per hour per year. This is only 115 miles per hour per hour. If this is applied continuously, it is 0.0016 g. If we use a 1 g higher acceleration rocket, then we can shut the engine down after three months, coast to within three months of the star, and then turn it on again to decelerate. Neither case—the 0.0016 g for 50 years, or 1 g for three months—sounds too bad, does it? Until you figure out what it means to the spacecraft. It requires a rocket exhaust velocity of about 20 million meters per second. This is orders of magnitude greater than anything ever even conceptually designed. A conventional chemical rocket has an exhaust velocity of 5,000 meters per second tops. Using nuclear rockets to power yourself by atomic bomb explosions might get you from 10,000 to 100,000 meters per second. Very advanced, and thus far theoretical, low-thrust ion-drive (magnetohydrodynamic) engines that use ionized plasma for propellant can theoretically give 100,000-to-200,000-meters-per-second exhaust velocities with accelerations near 0.001 g, but 20,000,000 meters per second?

The British Interplanetary Society has been a leader in providing a technical forum for interstellar studies. Several years ago the Society sponsored a design study for interstellar flight, which resulted in a design entitled Project Daedalus. The rocket in that study was a nuclear pulse rocket powered by very small hydrogen bombs exploded 250 times per second. Relativistic electron beams from the accelerator were used to initiate fusion reaction in deuterium and helium-3. (The helium-3 had to be mined on Jupiter!) Theoretically, if we could do all this, we could achieve exhaust velocities of about 100,000,000 meters per second with acceleration up to 1 g. This would do it. But it is a bit farfetched and even at that, it is impractical for general interstellar flight. A nuclear-electric system that could reach the nearest star in 400 years was recently analyzed at the Jet Propulsion Laboratory. It had a 1 million-kilowatt (gigawatt) nuclear-fission reactor. The radiation of heat from this reactor is a principal design challenge of any such vehicle. Even with the gigawatt power, nearly 500,000 kilograms (½ gigagram) initial mass and only 500 kilograms final mass, the top speed was only 0.012 times the speed of light—hence the 400-year trip time.

A matter–antimatter rocket would be the ultimate conventional rocket. It could produce an exhaust velocity equal to the speed of light. Antimatter is material, thus far discovered only as subatomic particles, which is the same as the corresponding matter particle but with opposite electric charge. The opposite of an electron (negative charge) is the antimatter particle called a positron. This is different from the proton, which also has opposite charge but is a very different particle—much more massive and, in fact, nuclear. Electrons and positrons exist outside atomic nuclei. When an electron and positron collide, or when any other matter–antimatter collision occurs, both are annihilated and energy (E) is created according to the famous formula, $E = mc^2$. That is, energy is created c^2 more powerful than m, the mass ($c^2 = 35$ billion miles squared per second squared). But where would we get enough antimatter to build such a rocket? We would need at least a few hundred kilograms for interstellar flight. Our capacity to make antimatter is now on the order of only 10 micrograms per year—worldwide! Some theoreticians think this can be improved a millionfold, but they admit that it might take an amount of power equal to the world's current output to do so.

This makes one thinking about interstellar flight realize that any rocket, the Project Daedalus rocket, or even the matter–antimatter rocket, will have an astronomical cost. The energy to power this thousand-ton machine with an exhaust velocity of 10^{18} meters per second is somewhere around 10^{18} kilowatt-hours—about 1 million years of U.S. power consumption! Even if that energy costs only 1 cent per kilowatt-hour, that's $10 million billion for fuel (10 thousand years or so of the U.S. national debt).

The fact is that it is not reasonable to consider interstellar flights with conventional rockets that have to carry fuel—any fuel. We should find a way to do interstellar flight without fuel or with fuel you can obtain while traveling. One other idea is to build a space ark and fill it with long-life robots or frozen people—or both—or to send forth whole colonies of explorers whose descendants will get there. But why? Instead of flying for, say, 10 thousand years, let's just wait a few hundred years until better technology is developed, and then go. The later vehicle would catch up to and pass the first one, anyway. This is, and for a while will continue to be, a point of discussion: Is it better to wait for new technology?

Since there is little point just now in debating the virtues and drawbacks of the ark-building concept, let's consider instead the possibility of using the one fuel we can obtain along the way: the hydrogen in the interstellar matter.

Robert Bussard first proposed this use of interstellar matter in his conceptual design for an interstellar ramjet engine. The idea is important and offers great possibilities, but it also presents large technical problems. For example, the ramjet would have to "inhale" enormous scoops (hundreds to thousands of cubic kilometers) of interstellar matter in order to get a usable amount of hydrogen, since only one or two atoms of hydrogen are present in each cubic centimeter in space—that is, 16 atoms per cubic inch. Then, when the gas has been collected, an enormous amount of power will be required to heat and accelerate it. Where will this power come from? It could be nuclear, or perhaps it could come from huge lasers in solar orbit—a possibility that needs further study. But even if we find a way to provide sufficient power, the technical challenge of designing and building the interstellar ramjet will be awesome.

And so, having eliminated the ark and the ramjet for now, let's consider the sail once again. It would provide a nearly perfect means of interstellar travel—except that it requires sunlight, and we will run out of that as we leave the solar system, a circumstance that could leave us becalmed, like a terrestrial sailboat on a windless ocean. We could, however, use a series of gravity-assist maneuvers to keep ourselves moving along. We could fly around Jupiter, for instance, then head in toward the sun, picking up a lot of speed with our sail. Then we could swing around and head out past Jupiter and Saturn, picking up even more speed. With a large sail and with a close fly-by of the sun, assuming good thermal protection, we could conceivably exit the solar system at a rate of 40 astronomical units* a year—in other words, we could travel through the whole solar system in about 1 year. Even at that great a speed, however, we would not reach the second star for 6,600 years! Not so good. Nevertheless, this use of gravity assist and solar sailing combined offers real promise for the use of solar sails even for traveling to the outer planets, which may extend the range of the interplanetary shuttle.

LASER SAILING

Let's further discuss the idea of using lasers instead of sunlight in interstellar space. Lasers are a means of producing light at

*An astronomical unit, or AU, is equal to the mean distance between Earth and the sun—in other words, 93 million miles (150 million kilometers).

a single wavelength from the quantum mechanical action in atoms. By stimulating atoms to change their level of energy together, lasers emit light at a wavelength determined by the amount of energy. With mirrors we can intensify this emission and amplify the light. The amplified light will stimulate more emission from the atoms making the total light beam even stronger. (For emissions at visible wavelengths of radiations the device is called a laser; if it is at far infrared wavelengths it is called a maser.) We could, of course, carry nuclear-powered lasers along with us and set them into orbit so that they would provide a light force for our sail, but that would be like carrying a nuclear-propulsion system—just another form of energy transfer and not advantageous. We can use lasers in orbit in the solar system, but they must be large. If we mean to match the power of the sun, we will need a billion-watt laser with an aperture. For sailing to the stars, we will need a laser 1,000 times more powerful than this, with apertures many kilometers in diameter. Developing these large apertures may not be difficult, since the necessary technology will probably be developed and lens systems can be devised for beam focusing. However, the high power levels are going to present a problem.

Robert Forward of Hughes Research Laboratories may know more than anyone else (on this planet, at least) about the possibilities for interstellar flight. He regularly collects and publishes a bibliography of all the literature on the subject. As a physicist who specializes in gravity analysis, he has come up with many ideas for interstellar designs and gravity-effect devices. He also writes science fiction. Forward has designed a fly-through sail vehicle that would get a 1-ton payload to the Alpha Centauri system in 40 years or even a bit less. The sail required is 3.6 kilometers in diameter (only about 2 miles—not that much larger than the JPL design). The sail is pure aluminum only 16 nanometers thick (0.016 micron, or 0.00063 mil, 0.63 millionths of an inch). Unlike "ordinary" sails, this pure aluminum sail will be constructed in space, and no plastic substrate will have to be carried along. Incidentally, we might be able to make the sail thinner, but that wouldn't help, because a thinner sail would be transparent, and the laser power would be lost as it passed through the sail.

This sail will reach a maximum velocity slightly greater than one-tenth the speed of light, and its acceleration will be 0.036 g. When this fast-moving vehicle reaches its destination, it will fly through the Alpha Centauri system at 36,000 kilometers per second (80 million miles per hour), which means we'll need a fast tape recorder and very quick reflexes (at this speed we would travel from the sun to Pluto in two days).

If we decide instead that we want to stop in the target system and stay there for a while, our job becomes a lot harder. Forward finds that for this stopover mission we will need a 7.2 trillion-watt laser with a 1,000-kilometer lens. The sail diameter will have to be 100 kilometers. Our maximum velocity will be one-fifth the speed of light, and the vehicle will have a decelerator to allow it to slow down when it reaches its target. This decelerator is basically a reflector larger than the main sail. When detached, it will speed ahead of the spacecraft and will reflect light back toward the main vehicle. The outer part of the sail can be this reflector stage. The main vehicle turns its small sail 180° toward the reflector and decelerates due to light pressure on its approach to the target star. This idea is illustrated on Figure 9.1.

This is a neat idea, although I think Jerry Pournelle and Larry Niven's approach is more realistic. In *The Mote in God's Eye*, they predict that if the civilized beings in the star system are interesting enough to stop and visit then they will know how to slow down our probe with their own technology.

This isn't the end of the subject, it's just the beginning. Robert Forward and Freeman Dyson have suggested other ideas such as perforating ultra-ultralight sails with holes smaller than the wavelength of light. This vehicle would be propelled by beamed microwaves (radio waves instead of light) and would serve as both a sail and an antenna. Microwave power is generated by masers instead of lasers, and it generates enormous accelerations and speeds.

Whatever the variations, laser or maser sailing is the most

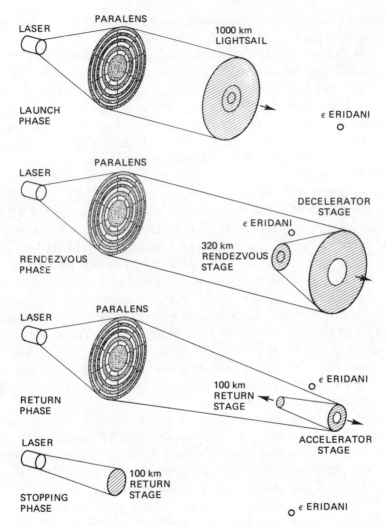

FIGURE 9.1 (1) The sail vehicle approaches the target star, Epsilon Eridani. (2) The outer part of the sail is detached. The inner part is turned around so that the nonreflective side faces the laser. The outer part moves ahead and the laser light is reflected back onto the inner part. The inner part then decelerates due to the reflected light pressure. (Robert Forward, Hughes Aircraft Company)

promising means of interstellar travel—not for us, unfortunately, but for our descendants.

SETI

There is, however, another way to "explore" the universe without using heavy and expensive fuel. It is SETI—the name is an acronym for Search for Extraterrestrial Intelligence. Through SETI programs, Earth scientists listen and search the sky for signals from hypothesized intelligent civilizations elsewhere in our galaxy. And whereas interstellar flight presents many technical limitations, SETI programs have only one: We can communicate only with those life forms that have evolved to the point where they can communicate. And please understand that I am not making a value judgment here. It is possible that less complex but far happier life forms exist in the universe, but if we are going to communicate with them, they will have to have the necessary communications technology. Just as we don't know about the probability of life forming on planets that are suitable to life forms, we do not know the probabilities of life evolving to a point where it wishes to, and is able to, communicate with other life forms. This limitation might not be quite so severe as it now seems. As our communications technology develops and we become willing to listen for "accidental" communications from life forms in the universe, we might find that many life forms make their presence known.

Interstellar communication has two overwhelming advantages: (1) we can do it *now*, and (2) it is much cheaper than interstellar flight.

Professor Paul Horowitz of Harvard University, the principal investigator on The Planetary Society's Project META (Megachannel Extraterrestrial Assay) search for extraterrestrial intelligence, has pointed out that an interstellar "telegram" costs approximately $1.00 a word (once we have the antenna and communications facilities), compared to the $100 million billion cost of fuel for interstellar flight. A 200-meter antenna receiving at a wavelength of 20 centimeters can receive a signal from a similar-size antenna 1,000 light-years away. The energy radiated in this system per word of information is less than 10 kilowatt hours—that is, about a dollar's worth. The same 200-meter antenna could also receive a signal from 30,000 light-years away—that is, from anywhere in the galaxy, if the transmitting civilization has a 6-kilometer antenna, probably in space, with a 10-megawatt transmitter. This is not such a wild assumption. We can envisage it for the planet Earth within 50 years. Radio searches will be the most economical way to search for the near future.

For the rest of the twentieth century, then, we should probably search for extraterrestrial intelligence by listening only. We should continue to develop designs for spacecraft that can complete the exploration of our own solar system. Having done that, we will take the first steps toward constructing a spacecraft to go to another star system. For that mission we can use the solar sail.

10

A RACE TO
THE MOON AND
OTHER ADVOCACY
FOR SOLAR
SAILING

Will we someday see solar sail craft racing to the moon? The Midi-Pyrenees, a
regional government in France, has agreed to sponsor such a race, and the Union
Pour la Promotion de la Propulsion Photonique (U3P) has already designed its entry.
(Illustration from *Science et Loisirs*)

Although the space agencies have not yet been able to look past their immediate needs and fund something so future-oriented as solar sailing, the concept continues to arouse the interest of the most talented and innovative thinkers everywhere. With no funding from their space agencies, groups of scientists in the U.S., Europe, the Soviet Union, and Japan continue to work on and promote solar sailing. Students write theses on the subject, amateurs form project groups, engineers propose test flights, and these dedicated people challenge one another to race to the moon. Even if solar sailing remains the stuff of which dreams are made, it will be the best of the dreamers who devote their attention to it, and they will do their best for space exploration.

I have seen the resonance that the solar sail excites in the public, in students, and in all those who are not immersed in bureaucratic quicksand. When we began the solar sail development program at the Jet Propulsion Laboratory, hundreds of articles about our work appeared in the world press, and we received countless letters offering support. A sailmaker in Hawaii, a retiree on a Caribbean island, the presidents of engineering companies, military personnel, and students volunteered to help us. Then, when the Halley Project was canceled, the solar sailing development ceased. The same thing happened in France. But from the Jet Propulsion Laboratory in California and some nearby aerospace companies, and from France's Centre National d'Etudes Spatiales and some French companies, young engineers, disgruntled at the bureaucrat's shortsightedness, banded together to develop the solar sail themselves. Thus were born the World Space Foundation (WSF) in the United States and the Union Pour la Promotion de la Propulsion Photonique in France. Both groups continue to design and develop solar sailing concepts.

The Foundation's concept is shown in Figure 10.1. Its designers have already fabricated a prototype sail of approximately 700 square meters and booms of 30 meters so that they can study fabrication and deployment techniques (see Figure 10.2). This is the only such sail actually built. They hope to work out a cooperative arrangement with NASA to fly a test vehicle on the shuttle in the 1980s. Their funding comes from individual and foundation donations, although they are seeking NASA support for their launch and industry donations for their materials and testing. The WSF's immediate goal is to fly a test vehicle; later they hope to use a sail to explore a nearby asteroid. The address for the World Space Foundation is P.O. Box Y, South Pasadena, CA, 91030-1000.

The Union Pour la Promotion de la Propulsion Photonique, known as U3P, is also seeking private sponsorship for its design and development of solar sailing spacecraft (see page 133). U3P

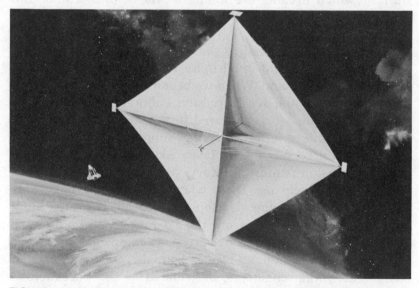

FIGURE 10.1 The World Space Foundation plans to fly a solar sail from the space shuttle sometime in the 1980s. Individuals and foundations have donated to the project, although the WSF is seeking NASA and industry support for materials and testing.

FIGURE 10.2 This prototype sail—developed by The World Space Foundation—was displayed to the public at Planetfest, '81, a celebration of the *Voyager 2* encounter with Saturn, hosted by The Planetary Society. (Photograph by Ken Tang)

has persuaded Midi-Pyrenees, a regional government in France, to sponsor a race to the moon and to offer a prize. The custom of using races and contests to spur technology development has a long and respectable history, and so the idea of a solar sailing race to the moon has wide appeal. The project is especially exciting to earthbound sailors, who test their boats and prove their seamanship in races. The address for U3P is 6 Rue des Remparts, Cologny, Venerque, 31120, Portet sur Garonne.

The moon race does offer one problem, however. The sail is best operated in deep space over interplanetary distances, not near Earth where it must make many maneuvers. The race, then, will be quite demanding for a "first" sail; it will require sophisticated navigation, guidance, attitude control, dynamic and stability control, maneuverability, and preciseness. Nevertheless, the race is technically possible, and it should stimulate even more widespread interest in solar sailing.

The rules of the race are simple:

1. All vehicles must start in geosynchronous orbit or lower. (Low Earth orbit would not be a good starting place, however, because of air drag and short maneuver time.)
2. All vehicles must use only photon pressure from sunlight on the Earth-to-moon transfer.
3. The first ship to go behind the moon wins.

Sailing vehicles of different designs—from very simple and small to immensely complex—can compete against one another in this race. I think special attention might be given to the smallest possible sail, which can most easily be controlled.

There are officially no entries yet, because U3P and the other

FIGURE 10.3 *The concept of solar sailing has captured imaginations around the world—especially those of enthusiastic and energetic young people. The Student Group for Astronautics at the Prague Observatory and Planetarium have designed a small heliogyro craft that may someday plow the space between planets. (Painting courtesy of Union Pour la Promotion de la Propulsion Photonique)*

groups have no sponsors. But the U3P, at least, has announced that it plans to go ahead with the race. The Foundation's sail will use the Moon as a mission target.

It's not just working scientists, however, who are excited about the sail. At an international meeting a few years ago, I was delighted when a Czechoslovakian teenager told me about a solar sailing study being conducted by the Student Group for Astronautics at the Prague Observatory and Planetarium. The students had designed a "minimum" solar sail spacecraft. They had decided on a small heliogyro because of the simpler design of that class and also because they could learn much about dynamical analysis by working on such a vehicle. In other words, these students had selected the heliogyro for the same reasons that JPL chose it for the Halley rendezvous attempt. An artist's concept of their design is shown in Figure 10.3. Interestingly, this student made his way to the U.S. where he is now in graduate school working on international space-exploration projects.

Student work is also going on in Great Britain, the United States, the USSR, and other nations. Engineers from British Aerospace have formed a private group known as the British Solar Sail Group. And there are solar sailing advocates in the general public, too. When we formed The Planetary Society in 1980, our first literature advocated "solar sailing trips to the planets." We have since become the largest space-interest group in the world, and we continue to encourage global interest in solar sailing. The Plantary Society address is 65 N. Catalina Av., Pasadena, CA 91106, USA.

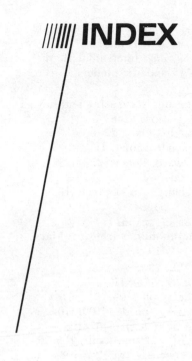

INDEX